Flocs in Water Treatment

Flocs in Water Treatment

David H. Bache and Ross Gregory

IWA PUBLISHING

Published by IWA Publishing, Republic - Export Building, 1 Clove Crescent, London E14 2BA, UK
Telephone: +44 (0) 20 7654 5500; Fax: +44 (0) 20 7654 5555; Email: publications@iwap.co.uk
Web: www.iwaponline.com

First published 2007

Cover design by www.designforpublishing.co.uk

British Library Cataloguing in Publication Data
A CIP catalogue record for this book is available from the British Library

Library of Congress Cataloging- in-Publication Data
A catalog record for this book is available from the Library of Congress

ISBN: 9781789065442

Contents

Preface

This book has been written in response to a perceived need to construct an information base relevant to the description of flocs and their characteristics as encountered in water treatment associated with water supply.

Although flocs play a critical role in solid-liquid separation, many aspects of their behaviour are barely understood. Much of our current understanding is based on the structure and dynamics of 'ideal' flocs, such as gained from the coagulation of monodisperse spherical particles under carefully controlled conditions. Flocs encountered in treatment works and within the natural environment display features of the 'ideal', but are far more complex in terms of structure and behaviour. They are difficult to measure and difficult to describe. One of the perturbing features affecting many studies within the literature is that they frequently fail to recognise these core problems. 'Real' flocs, *i.e.* those encountered in practice, are formed from a large variety of materials of different shapes and sizes, and are subject to changing hydrodynamic conditions and changing chemistry.

What does the book offer? To avoid any misunderstandings at the outset, it must be stated that the book does not represent the panacea for solving operational problems tied to chemical practices. Nor does it represent a fountain of knowledge concerning the role of flocs in treatment processes. Rather it reflects the first steps of a process in which many have sought to apply scientific discipline to what might be described as a sophisticated art *i.e.* running a waterworks. At present we can barely describe flocs as individuals, let alone contemplate the task of designing expert systems for supporting plant operations. What the book does offer is insight into the current state of knowledge and from this, readers can gain a deeper appreciation of the issues involved. In places this is fairly unpalatable, a pertinent example being the dispersion of hydrolysing coagulants. The effectiveness of a hydrolysing coagulant depends on the timescales of the underlying chemistry, the manner of coagulation, together with the timescales associated with mixing. The processes involved are complex and cannot be avoided. Hopefully the non-specialist reader will gain a greater appreciation of the issues and concepts involved, a step which (in the right circumstances) might ensure that appropriate mixing practices are put in place when using this type of chemical. Beyond that, the detail of the science can be tucked away.

For process scientists and engineers who have the task of trouble-shooting operational problems, the text is 'heady material' (as it has been for the authors), but should be useful in terms of explaining background concepts and identifying useful sources of information. For the scientific and engineering community concerned with development of the relevant science and technology, it is hoped that the book will provide supporting information for facilitating progress. It provides pointers to tasks that need to be addressed and indicates some of the blind alleys that are being pursued. Although the book has a strong focus on water treatment, the framework which is described applies to flocs and aggregates found in other circumstances. In this context Section 2 is particularly apt, as well as many parts of Section 1.

In terms of structure, the book has been divided into three parts. Section 1 provides insight into factors which affect floc formation and other background concepts. Its principal purpose is to provide an information base to service the needs of the book. Particular attention is given to the chemistry of hydrolysing coagulants in common use, because of their impact on coagulation and floc properties. Some of the material is fairly standard and well known, whereas other material has been drawn from recent literature. Section 2 provides a framework for describing floc properties. At first sight it may appear rigorous and comprehensive. In places this may be true, but the reality is closer to the structure of flocs, *i.e.* full of holes. Section 3 addresses three major themes, the first of which focuses on a number of problems that are encountered when measuring flocs. As indicated earlier, the measurement of flocs is not straightforward. Indeed much of the contradictory information about flocs that appears in the literature can be attributed to the vagaries of measurement. We have omitted many aspects *e.g.* measurement of rheological properties, these being addressed in specialist texts. The second theme in Section 3 is more straightforward – a commentary on the factors that control floc properties. This draws on Sections 1 and 2, the text being provided with many cross-references to service the needs of the reader. The final theme concerns flocs in treatment processes. Material has been selected to describe particular themes and avoids the task of collating the vast literature surrounding individual processes. We have sought to include more general information which may be of assistance to those dealing with plant operation.

The book has aimed to provide a balance between explanation and the demands of the supporting science. One of the great difficulties during the preparation of the text has been the task of coping with contradictory information (aluminium polymerisation and its role in coagulation is a good example, another being the diversity of values of fractal dimension reported in the literature for similar materials). In places we have placed a particular interpretation on data – others may not share this view. Some topics in which there is much uncertainty have required 'delicate footwork' in terms of what is stated in the text. Overall, what we do hope is that the book will provide a source of information and inspiration to designers and practitioners as well as to the academic community in terms of advancing the understanding of flocs and their role in water treatment.

DHB, RG 2007

Section 1 Background Concepts

1.1 INTRODUCTION

1.1.1 Water quality perspective

The composition of water is moulded by its history. As rain passes through the atmosphere, it dissolves gases and traps aerosols. On reaching the earth's surface, it either evaporates or continues its flow in various guises of runoff. The energy associated with surface flow determines its erosive potential - resulting in the suspension of clays and a wide variety of particulate matter. In the form of sub-surface flows, there is greater contact time with crustal components, this allowing water's capacity as a solvent to increase. The gains depend on factors such as solubility and crustal abundance. One may see the appearance of elements such as Fe, Mn and As, as well as those which strongly influence the alkalinity and pH.

Table 1.1. Average raw water quality from different source types (Hall, 1997).

Parameter	Upland reservoir	Upland river	Groundwater	Lowland reservoir	Lowland river
Colour (°H)	36.4	22.9	5.5	11.9	23.6
Turbidity (NTU)	2.9	2.7	2.2	3.6	7.0
Manganese (mg/l)	0.091	0.029	0.034	0.053	0.064
Iron (mg/l)	0.33	0.18	0.28	0.17	0.42
Hardness (mg $CaCO_3$/l)	35.6	50.9	291.5	208.8	146.0
pH	6.65	6.65	7.17	8.0	7.66
Coliforms (No./100ml)	72.2	741.0	38.4	282.1	2.2×10^4

ISBN: 9781789065442. Published by IWA Publishing, London, UK.

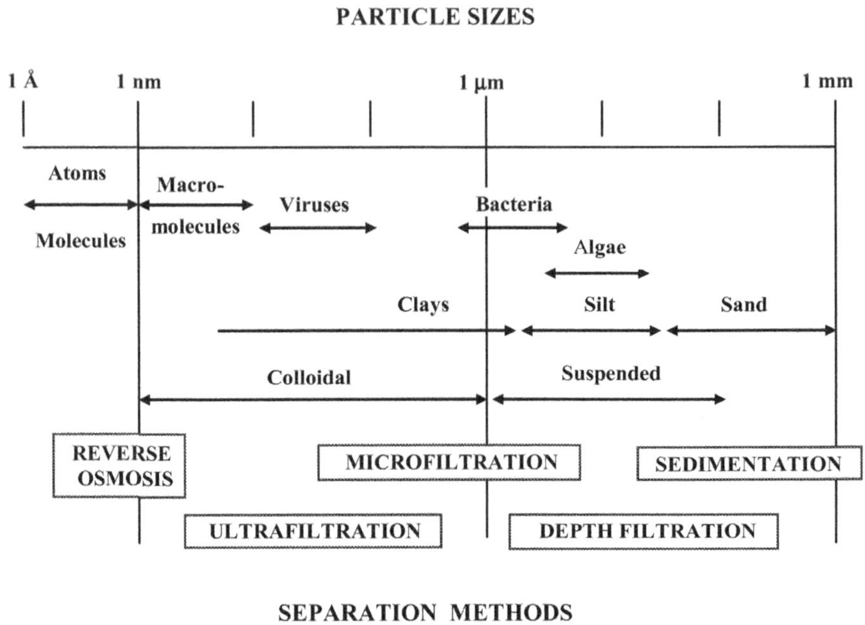

Figure 1.1. Typical size ranges for particulates found in water together with ranges associated with separation processes (based on Elimelech *et al.*, 1995).

Beside the naturally occurring materials, there are contributions from industrial effluents, sewage and farmland runoff, each adding to the cocktail of materials which are present. Water will also contain a plethora of living matter ranging from microorganisms to small plants and animals along with their residues. Table 1.1 illustrates some common traits associated with different types of source.

While information of the type shown in Table 1.1 is useful for highlighting typical features of a water source, it lacks detail of the factors which affect its treatability. For this, one must look more closely at the nature of the materials held by water. For example, colour is strongly associated with humic acid and fulvic acid. These materials have different molecular weights and sizes and are not equivalent in terms of their response to coagulation and separation in conventional treatment. Coliforms can be considered as a specific type of particle which can be filtered out rather than being inactivated by a disinfectant. Turbidity is most strongly influenced by particles in the size range 0.1-1µm (Huber and Frost, 1998). Elements such as Mn, Fe and hardness may also show their presence in particulate form, depending on the pH. Particles are ubiquitous in natural waters and often exist in concentrations of 10^7 to 10^8 per litre (Stumm, 1992). Generally they occupy a continuous size range from molecular dimensions through to a few mm as evident in Figure 1.1.

The kinematics of particles and their behaviour in solid-liquid separation processes is strongly influenced by their size and surface properties. Figure 1.1 indicates that particles greater than a few μm can be removed by the combination of sedimentation and filtration, whereas for smaller sizes one either has to resort to different forms of membrane filtration or to promote aggregation so that separation technology for larger particles can be exploited. As particle sizes get smaller, their surface to volume ratio increases – implying that interfacial properties play an increasing role in particle dynamics.

1.1.2 Surface characteristics

The wide range of materials found in water would seem to imply a great diversity in the character of their interfaces. Fortunately there are many common patterns of behaviour which allow surface characteristics to be described in a systematic way (Stumm, 1992).

Atoms, molecules and ions exert forces on each other, a factor which does not stop at their interfaces. Many solid surfaces contain ionisable functional groups such as the hydroxyl (OH) and carboxyl (COOH) which serve as electron donors in surface coordination chemistry. The surface charge depends on the degree of ionisation, this being influenced by the pH of the suspending medium. Surface charges also arise from imperfections in the lattice structure and are modified by adsorption. The existence of surface charge gives rise to electric interactions which extend over much greater distances than chemical bonds. Fluctuations in the electromagnetic fields around atoms and molecules induce temporary dipoles whose interaction yield van der Waals forces. These attractive forces play an important role in adsorption and inter-particle bonding. As with electric interactions, van der Waals forces are long range, implying that they are significant over distances which are large relative to atomic and molecular dimensions. For reference purposes it is useful to note that a water molecule has a diameter of about 0.3 nm.

Beside interactions involving coordinative bonds and electric fields, particle behaviour is also influenced by surface-solvent interactions - surfaces being classed as hydrophilic and hydrophobic, or *amphipathic*, in which both classes of behaviour exist.

Hydrophilic materials are water soluble. The properties are conferred by the presence of polar and ionic groups and their affinity for water, with the consequence that the surface may become hydrated. Water molecules are polar, the charge separation creating a dipole. In bulk water, transient structures exist between water molecules because of hydrogen bonding – sometimes tending towards a hexagonal structure, such as found in ice crystals. When an ion is present, this alters the hydrogen bond network. A water molecule tends to reorient so that its polarised charge concentration faces the opposite charge of the ion. As the water molecules orient themselves toward the ion, they break the hydrogen bonds to their nearest neighbours (see Figure 1.2). The group of water molecules oriented about an ion is called a hydration shell. The orientation of molecules in the hydration shell results in a net charge on the outside of the shell, the charge having the same sign as the ion. Charges on the outside of the hydration shell tend to orient water molecules in their vicinity, leading to the formation of a second hydration shell.

If a particle surface is charged, it will tend to hydrate in similar same way to the ion with dipoles oriented towards or away from the surface depending on its charge. Under the influence of electric fields the ordering of water molecules close to the surface leads to higher values of dynamic viscosity and lower values of dielectric constant when compared with those found in bulk water.

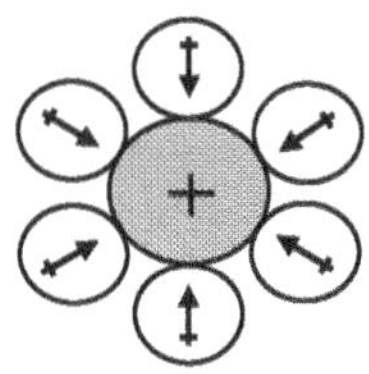

Figure 1.2. Isolated cation surrounded by oriented dipoles in first hydration layer.

When a surface has no polar or ionic groups or hydrogen bonding sites, there is no affinity for water and the surface is described as *hydrophobic*. Hydrophobic substances are sparingly soluble in water and tend to be adsorbed on relatively non-polar surfaces. The adsorption is not founded on attraction between the material and the surface. Instead, it stems from the natural tendency of water molecules to associate with each other rather than with a hydrophobic substance; this leads to a thermodynamic 'rejection' of a hydrophobic particle from the bulk water continuum.

Many organic substances, such as fatty acids and detergents contain both hydrophobic parts and hydrophilic groups. This amphipathic characteristic also applies to humic and fulvic acids because they contain various types of phenolic and carboxylic ionisable hydrophilic groups as well as aromatic and aliphatic moieties which impart hydrophobic properties. If the molecular weight of organic dipoles and organic ions is sufficiently high, the hydrophobic tendency derived from the hydrocarbon building blocks tends to dominate their overall behaviour and may result in hydrophobic bonding.

When addressing the problems faced in coagulation, one is always going to be dealing with factors which stem from the types of the materials involved and their physico-chemistry. An example of a classification scheme which attempts to identify compositional features which affect the behaviour of coagulation admixtures is shown in Table 1.2.

Table 1.2. Compositional features affecting coagulation (adapted from Dymaczewski *et al.*, 1997)

Type of coagulated admixture	**Compositional Feature**					
	Colloid particle	Macro-molecular colloid	Hydro-phobic colloid	Hydro-philic colloid	Mineral	Organic
Clay particles	+		+		+	
Humic substances		+	+	+		+
Proteins, carbohydrates		+		+		+
Aluminium and iron hydroxides	+			+	+	
Calcium carbonate, Magnesium hydroxide	+		+		+	
Inorganic flocculants (bentonite, fullers earth)	+		+		+	
Activated silica	+		+		+	
Polyelectrolytes		+		+		+

1.2 COLLOIDS

1.2.1 General concepts

Conventionally, colloids refer to particles in the size range 1 nm to 1 μm. At the lower end of this scale, the size and surface properties are sufficient to define a distinct interface between the dispersed phase and the dispersion medium. In this section, detail is provided of the force fields which control the interaction between particles. Those of immediate concern are the longer range forces.

In an electrolyte, the charge on a particle surface is balanced by an equivalent number of oppositely charged counterions in solution (Figure 1.3). The counterions are subject to two opposing influences: electrostatic attraction which promotes the clustering of counterions near the surface and thermally driven diffusive motions which serve to reduce concentration gradients and clustering. The charge on a particle and its sheath of counterions constitute an *electrical double layer* (EDL).

If two particles approach one another, the sheaths of counterions interact and repel each other. However, if the particles can get sufficiently close to one another, the electrical repulsive forces will be overcome by attractive van der Waals forces and the particles will coalesce. For situations in which the repulsive forces dominate the inter-particle kinetics, the particles cannot coalesce and the suspension will remain in a disperse state. In kinetic terms, a dispersion is described as being in a *stable* state, in the sense that if coalescence does take place, it is very slow. One of the key objectives in coagulation is to circumvent this problem so that the suspension can be *destabilised* and allow aggregation to take place more rapidly.

The framework for describing the electrical and van der Waals interactions was developed independently by Derjaguin and Landau (1941) and Verwey and Overbeek (1948) and is known as '*DLVO theory*'. However, there are many other force fields which need to be taken into account, such as those stemming from particle-solvent interactions. Following the nomenclature used in Elimelech *et al.* (1995), these additional forces (*i.e.* beyond van der Waals and EDL interactions) are termed '*non-DLVO forces*'.

The impact of the force fields is often represented by the interaction energy which measures the amount of work expended (*i.e.* potential energy) in bringing the particles towards each other. By convention, this is positive for energy expended in overcoming repulsive forces and negative for attractive forces. When dealing with a combination of forces, the total potential (V_T) can be estimated by the summation of the component energy terms *i.e.*

$$V_T = V_B + V_A + V_R + V_S \tag{1.1}$$

in which V_B, V_A, V_R and V_S refer to the Born, van der Waals, electrical double layer interactions and structural energy terms respectively. Other interactions can also be added to this list and will be considered in due course. The Born forces are extremely short range repulsive forces which arise between atoms as their electron shells penetrate each other. Structural forces describe the interactions arising specifically from the adsorption of solvent, surfactant or macromolecules at the interface.

1.2.2 Van der Waals interaction

Van der Waals forces are always attractive between particles of the same kind. Hamaker (1937) derived equations for these forces on the basis of the additivity of van der Waals energy between pairs of atoms or molecules. The interaction between particles is obtained by summing the interactions of all molecules in one particle with all the molecules in the other.

For the case of two identical spheres of radius a, with interfaces at the distance of closest approach separated by the distance h, the interaction energy (V_A) when the particles are close together (h << a) is given by:

$$V_A = -\frac{A_H a}{12h} \tag{1.2}$$

in which A_H is the effective Hamaker constant for the particle-liquid combination. For practically all aqueous dispersions, Gregory (1989) reports Hamakar constants in the range 0.3 to 10 x 10^{-20} J. Dense materials have values toward the upper end of this range, whereas low density matter, such as biological materials, have relatively low values. More complex expressions for V_A arise for particles of different size and differing media and in circumstances in which the condition h << a does not hold. The salient features of Equation (1.2) are the proportionality between V_A and particle size and the inverse dependence on the interparticle spacing (h). Van der Waals forces act over a range which is comparable with particle size, but are much reduced by retardation (*e.g.* Elimelech *et al.*, 1995) when the spacing exceeds about 10 nm (Shaw, 1992).

1.2.3 Electrical double layer

In Section 1.2.1 it was noted that the charge on the colloid attracts counterions in solution and co ions are repelled. The resultant distribution is often described in terms of the Stern model, in which the distribution of counterions is subdivided into two zones as shown in Figure 1.3. The inner zone (Stern layer) consists of a compact layer of counterions adsorbed at the surface. It may also contain specifically adsorbed ions. Overall the Stern layer is about 0.3 to 0.5 nm in extent, corresponding to the diameter of a hydrated ion. Beyond the Stern layer is the diffuse layer in which ions are distributed according to the influence of electric forces and random thermal motion. The outer edge of the Stern layer (distance δ) more or less coincides with the plane of shear. This implies that ions within the Stern layer move with the colloid, whereas those further from the surface have greater freedom of movement. Because the plane of shear defines the effective hydrodynamic surface, the electrical potential at this position (*i.e.* the *zeta potential,* ζ) plays a critical role in electrokinetics. The zeta potential differs from the surface potential as illustrated in Figure 1.3. Provided the zeta potential is sufficiently low (say < 25 mV in a 1:1 electrolyte) the electrical potential (ψ) follows the distribution:

$$\psi = \psi_\delta \exp(-\kappa(r-\delta)) \tag{1.3}$$

in which ψ_δ ($\approx \zeta$) is the potential at the edge of the Stern layer, r the distance from the particle surface and κ is the Debye-Huckel parameter. The Debye-Huckle parameter determines the extent of the counterions charge in solution (or the diffuse layer). The value $1/\kappa$ defines the Debye length and is used to represent the 'thickness' of the diffuse layer. Its value can be gained via the definition:

$$\kappa = \left(\frac{2000 F^2 I}{\varepsilon_r \varepsilon_o RT} \right)^{1/2} \text{ in units of m}^{-1} \tag{1.4}$$

in which F is Faraday's constant, R the gas constant, T the absolute temperature (K), ε_r the dielectric constant (also known as the relative permittivity), ε_o the permittivity in vacuum and I the ionic strength ($=\frac{1}{2} \Sigma\, c_i\, z_i^2$ where c_i is the ionic concentration in mol l^{-1} and z_i the associated ionic charge). At 25° C with $\varepsilon_r = 78.5$ for water, equation (1.4) reduces to:

$$\kappa = 3.288 I^{1/2} \text{ (nm}^{-1}\text{)} \tag{1.5}$$

In 10^{-3} M 1:1 aqueous electrolyte solution the Debye length is about 10 nm, whereas for seawater with much greater ionic strength $\kappa^{-1} \approx 0.4$ nm.

When the zeta potential is low and $\kappa a >> 1$, the effective charge density (C m^{-2}) at the plane of shear (balancing the charge held within the diffuse layer) is given by the approximation:

$$\sigma_\delta = \varepsilon \kappa^{-1} \psi_\delta \tag{1.6}$$

where ε ($= \varepsilon_r \varepsilon_o$) refers to the permittivity. This shows that the parameters σ_δ and ψ_δ (or ζ) are virtually synonymous in terms of describing the electrical state of the effective surface (at δ).

1.2.4 Double layer interaction energy

The calculation of the interaction energy, V_R, resulting from the overlapping of two spherical particles is complex. No exact analytical expressions can be given and recourse must be made to approximation solutions, such as summarised in Elimelech *et al.* (1995).

When two surfaces approach it is likely that surface charge or surface potential will be modified. One is also concerned with the degree of overlap of the double layers.

For identical particles with small values of surface potential satisfying the conditions h << a and κa > 5:

$$V_R = 2\pi\varepsilon a\zeta^2 \exp(-\kappa h) \tag{1.7}$$

This shows that the repulsive energy is proportional to the radius and is strongly dependent on the magnitude of the zeta potential.

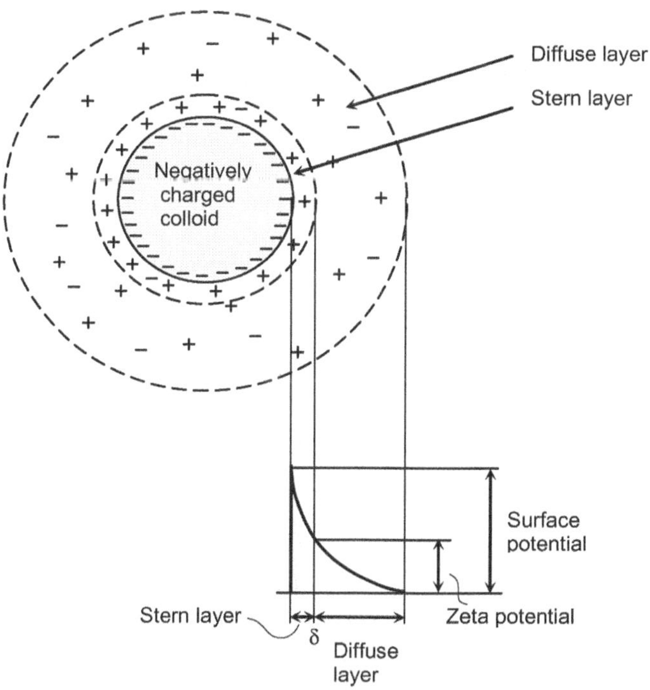

Figure 1.3. Electrical double layer of a negatively charged colloid (based on Sawyer *et al.* 1994).

1.2.5 Combined interaction

When the van der Waals and EDL interaction energies shown by Equations (1.2) and (1.7) are combined (as in the DLVO approach), the total potential is specified by:

$$V_T = 2\pi\varepsilon a\zeta^2 \exp(-\kappa h) - A_H a/(12h) \tag{1.8}$$

This is subject to restrictions of the component terms and is only a partial description of V_T. Nevertheless it is quite useful, because it illustrates the basic influences on colloid stability. When h is very small, V_R tends to a finite value whereas the magnitude of V_A becomes large showing that attractive forces dominate at small distances. At larger interface distances, and in cases where V_R exceeds V_A, the total potential passes through a maximum known as the energy barrier (Figure 1.4) At still greater spacings, the EDL interaction diminishes more rapidly than the van der Waals, with the result that van der Waals forces (albeit very weak) will tend to dominate at larger distances. This may give rise to a secondary minimum at a spacing of roughly $7\kappa^{-1}$ (Hunter, 2001).

Beside V_A and V_R, non-DLVO interactions need to be taken into account. Inspection of Equation (1.2) shows that when $h \rightarrow 0$, $V_A \rightarrow \infty$. This does not occur because Born forces come into play and strongly influence the depth and position the primary minimum; the latter occurs at a spacing of about 0.1-0.2 nm (Elimelech *et al.*, 1995).

The DLVO theory treats an ion as a point charge and ignores both its size and the energy involved in moving aside other ions, or any other effect which the ion will have on the local structure of the solvent or the distribution of other ions. This form of energy comes under the term V_S shown in Equation (1.1). V_S accounts for the energy expenditure which accompanies the displacement of water molecules if particles are to come into close contact. It also covers steric interactions.

1.2.6 Hydration and hydrophobic interactions

If particles are to come into close contact so that they 'touch', this requires the displacement of water molecules surrounding the particles. This is when the hydrophilic or hydrophobic character of the surface exerts an influence. Greater amounts of energy are required to displace water molecules held in the hydration layers surrounding a hydrophilic colloid when compared to the energy demand for displacing water molecules adjacent to a hydrophobic surface.

Interactions of the type described above can be examined in various ways, and it is useful to refer to the description provided in Hunter (2001, p 614). One of the consequences of a surface-solvent interaction is to cause variations in the number density of water molecules away from the surface. This causes variations in the associated osmotic pressure away from the surface, thereby inducing a force field. For a hydrophilic surface, hydration gives rise to a repulsive force (*hydration force*) whereas a hydrophobic surface induces attractive forces as a result of these phenomena. Both types of interaction are significant for separations of up to about 5 nm. Both decay exponentially with decay lengths typically of the order of 1 nm (Hunter, 2001).

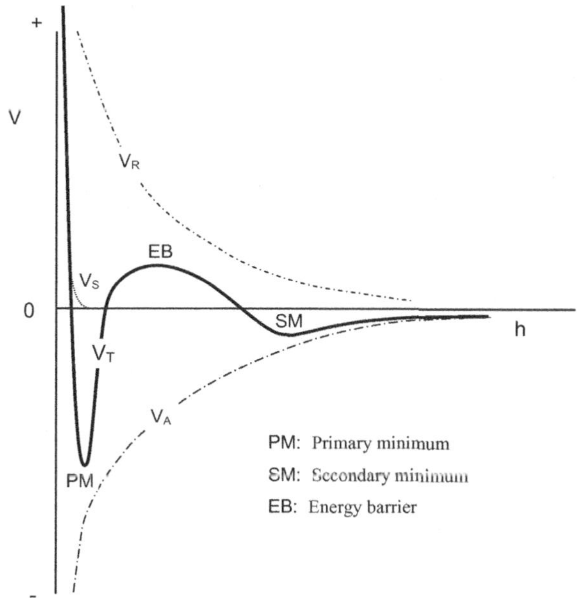

Figure 1.4. Schematic trend of total potential energy of interaction V_T showing salient features with V_S as the energy of repulsion due to a solvent layer.

1.2.7 Steric stabilisation

Steric stabilisation is a name used to describe several different stabilising mechanisms involving adsorbed macromolecules, such as long-chained polymers. These can be anchored at the surface and extend into the aqueous phase in the form shown Figure 1.5. When two such particles come into close contact, the number of chain-chain contacts increase at the expense of chain-solvent interactions, resulting in repulsion between the adsorbed layers.

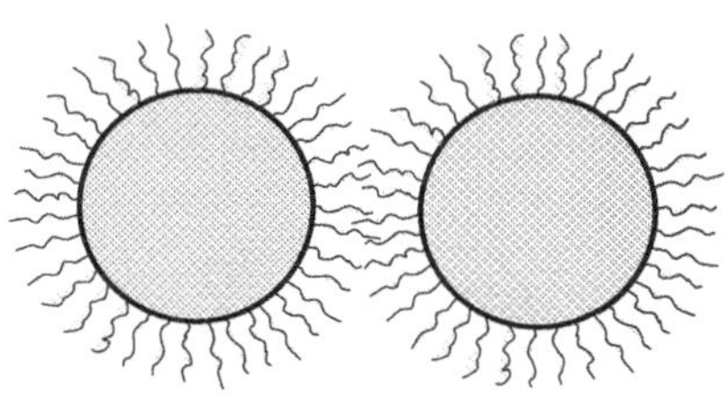

Figure 1.5. Steric stabilisation by adsorbed polymers, chain interactions preventing close particle contact.

Adsorbed macromolecules also alter the van der Waals forces between the particles in the following ways. First, the layers introduce a new medium with its own Hamaker constant (Section 1.3.2). Second, the combined thickness of the adsorbed layers acts as a physical barrier which keeps the particles apart, thus reducing their mutual interaction. The two influences can be brought together in a revised version of Equation (1.2) (see Sonntag and Strenge, 1987). Of the two influences, Elimelech *et al.* (1995) indicate that the displacement factor is the more important.

1.2.8 Polymer bridging

Polymers are widely used in water treatment, occasionally serving as a primary coagulant or more often used as a means of enhancing floc strength or dewaterability. They come in many different forms depending on their molecular weight, charge type *i.e.* cationic, anionic or non-ionic and charge density. For the bridging mechanism to work in the role of a primary coagulant, the basic requirement is that the polymer adsorbs on one particle and is of sufficient length to straddle the distance over which repulsion forces are effective and to adsorb onto another. Adsorption may be in the form of coulombic electrostatic reactions or arising from van der Waals attraction and hydrophobic bonding. Where the particle and polymer are of opposite sign, adsorption by electrostatic attraction will reduce the net charge on the particle and improve the prospects of destabilisation as a result of reductions in the energy barrier. Generally the most effective bridging flocculation is found with linear polymers of very high molecular weight (several million), so that extended loops and tails can form and increase the probability of attachment with other particles.

Where the polymer has a high charge density and the charge density on the surface is low, the polymer chain tends to form electrostatic bonds at intermittent positions along its length and at intermittent positions across the surface of the particle. This forms a patchwork across the particle in which some regions are dominated by the unbalanced polymer charge and other areas in which the original charges remain. If patches of opposite charge on the adjacent particles come into contact, strong coulombic attraction may prevail and hold the particles together.

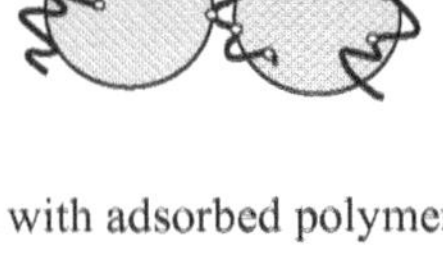

Figure 1.6. Bridging flocculation with adsorbed polymer.

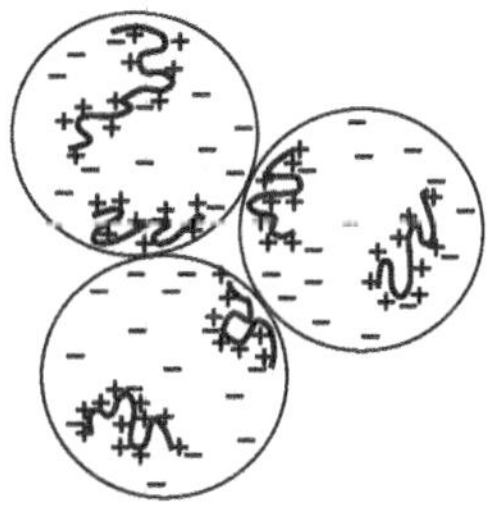

Figure 1.7. Bridging via electrostatic patch interactions.

1.2.9 Electrostatic bridging

Electrostatic bridging was a term employed in Bache *et al.* (1999) to explain the relative strength of flocs formed from a series of suspensions of similar particle size, but differing zeta potential (see zero dose in Figure 1.8). The suspensions were each coagulated with aluminium sulphate at the same pH. Floc strength data at 5 mg Al/l showed the relative strength of the flocs followed the same order as the ranking of the charge on the primary particles - the weakest flocs (latex) associated with the least charge and the strongest flocs (Minusil 5) associated primary particles carrying the largest charge. When the Al dose was 5 mg/l, Figure 1.8 shows that each suspension has the same zeta potential. N.B. the pattern of response to dose in Figure 1.8 indicates the formation and existence of a surface coating formed as a result of specific adsorption of aluminium hydrolysis products. When coating is complete, it presents a new surface which controls the zeta potential. The common zeta potential (above 4 mg/l) for each suspension suggests that the coating is formed from the same material in all cases.

Given that that the coated particles were of similar size and have the same zeta potential, differences in strength cannot be explained by the DLVO model (see Bache *et al.*, 1999). Rather it appears that bonding was being promoted by attraction between the negative charges of the primary particles and positively charged deposit located *between* the particles. This is the basis of an electrostatic bridge shown schematically in Figure 1.10.

On the assumption that the bonding behaves in the same way as two point charges q_1, q_2 separated by a distance r, the bond strength (F) is specified by:

$$F = \frac{1}{\varepsilon}\frac{q_1 q_2}{r^2} \tag{1.9}$$

where ε refers to the permittivity of the dielectric medium *e.g.* water. The implications of this type of bonding can be examined in the context of

Figures 1.8 - 1.10. If the charges q_1 are q_2 are associated with the zeta potentials of the uncoated primary particles and the coated layer respectively, then the strength of the interaction force is scaled by the magnitude of the product q_1q_2. Because all coated particles have the same zeta potential (when strength was measured), one can assume that q_2 is identical for each suspension. This implies that the magnitude of the product q_1q_2 is controlled by the primary charge 'q_1', thus explaining the observed ranking in strength. The precise nature of this form of bonding has still to be explored, but evidence of its existence is compelling.

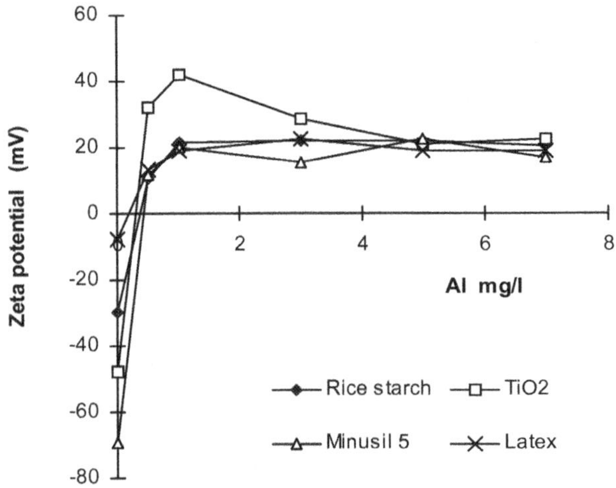

Figure 1.8. Dependence of zeta potential on the Al dose at pH 6.3 and controlled ionic strength for four suspensions (from Bache *et al.*, 1999).

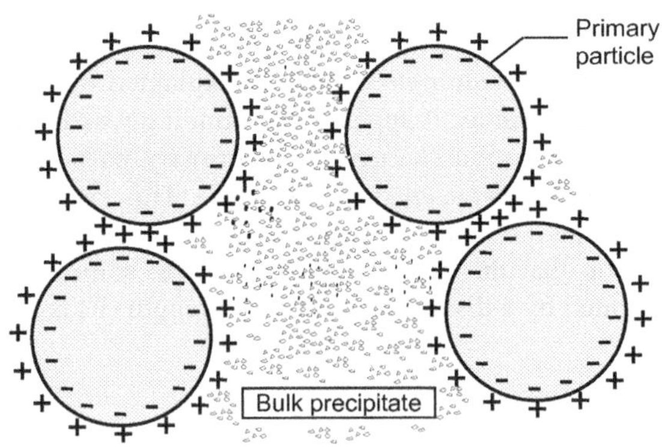

Figure 1.9. Conceptual model of floc matrix associated with sweep floc (from Bache *et al.*, 1999).

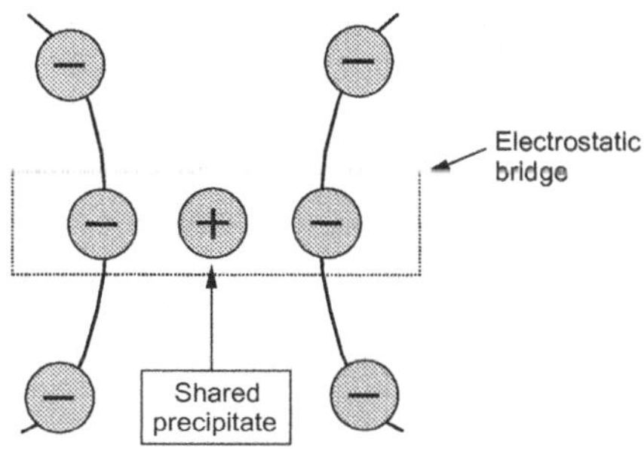

Figure 1.10. Schematic view of an electrostatic bridge based on point charges (from Bache *et al.*, 1999).

1.2.10 Stability and binding

From the brief resumé, it is apparent that many factors influence particle interactions. In the first instance, they provide insight into the causes of dispersion stability and of potential routes for destabilising the suspension. Second, they provide insight into a suite of binding mechanisms which hold particles and flocs together.

If a large energy barrier exists, particles will be unable to coalesce unless the effects of the barrier can be circumvented. At the other extreme, if there is no potential barrier, particles will be drawn together to a spacing commensurate with the lowest possible energy state (primary minimum). If the particles 'touch' *i.e.* $h \approx 0$, the energy minimum will be deep and the particles held together by strong attraction forces. This contrasts with the case of adsorbed layers whose presence keep the original particle surfaces further apart (as described in Section 1.2.7); this alters both the position and magnitude of the primary minimum, such that the binding forces are weaker than those described above.

Energy barriers can be modified or circumvented by coagulation practices (Section 1.3.2). If conditions are such that a small barrier exists, this may not hinder the coagulation process because particles may possess sufficient kinetic energy to overcome the influence of the barrier. When particles have diameters less than about 10 μm most of their kinetic energy (under typical mixing conditions) is dominated by the thermal contribution. In the absence of external forces, all suspended particles (regardless of their size) have the same translational kinetic energy (1.5kT) in which k is the Boltzmann constant and T the absolute temperature. To assess whether a barrier is 'large' or 'small' relative to available kinetic energy for this domain of particle sizes, it is useful to represent the interaction energy in the form of the ratio V_T/kT. When V_T is a few kT there is generally sufficient thermal kinetic energy to enable coagulation to progress. On the other hand if $V_T/kT > 20$ the system is essentially stable (see also Section 1.3.3.2). Similarly, if particles are held together in a secondary minimum with a depth of a few kT, there is strong likelihood that the bond will be broken at some instant. In contrast, when particles are bonded at the primary minimum there is little prospect of breaking the bond *i.e.* coagulation is irreversible.

In the classical regime of *slow aggregation* (*i.e.* with an energy barrier) retention in the secondary minimum becomes pronounced when either the ionic strength of the solution is high (*e.g.* 0.1 M for particles of 150 nm radius) or particles are large (≥ 350 nm radius for an ionic strength of 0.01 M (Behrens and Borkovec, 2000). It is also favoured by materials with a large Hamakar constant. When particles are small (say < 10 nm), a secondary minimum is never deep enough to serve any useful value (Shaw, 1992). When particles are bonded at secondary minima, the bonds are relatively loose *i.e.* flexible. Associated aggregates will have greater water content than clusters founded on bonds at the primary minimum.

1.3 FLOC FORMATION

1.3.1 Introduction

Floc growth can only take place if a series of conditions are satisfied. First, there must be a supply of building material; this can be gained by direct nucleation on the particle surface, adsorption of materials held in suspension, or from inter-particle aggregation. Second, the thermodynamic state must be conducive to bond formation. If a colloidal suspension is thermodynamically stable, it implies that it must be destabilised to allow aggregation to either proceed or to progress more rapidly. The supply rate of material for supporting growth depends on its abundance (a function of concentration) and on transport factors.

With respect to terminology, the terms coagulation and flocculation are often used somewhat loosely to describe the overall process of aggregation. Here, we regard coagulation as the process associated with destabilisation *i.e.* overcoming the factors which contribute to the stability of the suspension. Flocculation generally refers to the transport phase in which particles come into contact as a result of relative motion. If a suspension has been sufficiently destabilised, collisions *may* result in bond formation and growth in particle size.

1.3.2 Mechanisms of coagulation

Destabilisation can be achieved in various ways, the mechanisms depending on the type of coagulant, its concentration, the nature of the colloids and the prevailing chemistry. The basic task is to overcome the effects of the energy barrier. There exist two principal routes for achieving this task. In the first, destabilisation is promoted by reducing the energy barrier, this providing conditions in which van der Waals or other adhesive forces can dominate. Barrier reductions can be induced by reducing the spatial extent of the double layer *i.e. double layer compression* or by reducing the surface potential as a result of changes in the surface chemistry. In the second group, a significant energy barrier may persist, but its kinetic impact can be circumvented by mechanical bridging or by entrapment within a precipitate. In practice, a combination of destabilisation mechanisms may be active.

1.3.2.1 Double layer compression

When an indifferent electrolyte (*e.g.* a solution of NaCl) is added to a stable colloidal suspension there are no significant chemical interactions and the effects are predominantly electrostatic. Equation (1.4) shows that the Debye length ($1/\kappa$) is a function of the ionic strength (I) *i.e.* $1/\kappa \propto I^{-1/2}$, implying that increases in 'I' will lead to reductions in the thickness of the double layer. The associated increase in the value of the Debye-Huckle parameter (κ) leads to a decrease in the magnitude of the repulsive potential (Equation 1.7). In consequence, the total potential $V_T = V_R + V_A$ is reduced. Ultimately the energy barrier can be reduced to zero (*i.e.* $V_T = 0$) when the ionic strength is sufficiently great. This condition is often identified by a *critical coagulant concentration* (c.c.c.) at which rapid aggregation occurs; its value is independent of the concentration of the colloids.

By measuring the c.c.c. of an indifferent electrolyte, useful information can be gained about some the factors controlling stability. However, the coagulants used in normal practice do not behave as indifferent electrolytes because of their influence on the surface chemistry. Although double layer compression exists, it can usually be disregarded in water treatment. It becomes significant in situations such as estuaries or particular industrial applications in which the ionic strength is large.

1.3.2.2 Adsorption and charge neutralisation (ACN)

Many types of particle found in raw water are negatively charged at typical pH and will obviously attract positively charged materials as a result of long-range electrical interactions. Ions (of whatever kind) that enter the Stern or Stern-like layer to be held by *strong* surface forces are said to be *specifically adsorbed* and may influence the surface charge. Cations arising from the hydrolysis of metal coagulants (or cationic polymers) are prime candidates for specific adsorption. The impact of coagulant dose on the zeta potential (or surface charge) is illustrated in Figure 1.8. As the charge and energy barrier are reduced, coagulation progresses (trend AA′ in Figure 1.11). When the surface charge has been neutralised, particle interactions are dominated by van der Waals forces - allowing rapid coagulation. This state more or less coincides with the minimum shown in Figure 1.11. With increasing dose and continued adsorption, charge reversal can take place (see Figure 1.8) increasing the energy barrier and causing the suspension to restabilise (trend B′B in Figure 1.11).

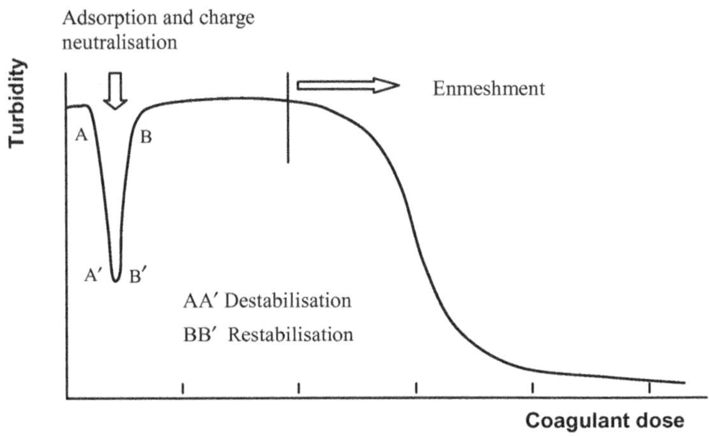

Figure 1.11. Schematic dependence of turbidity on coagulant dose showing regions of destabilisation.

In principle, one should be able to predict the impact of a coagulant on the electrokinetic state. A very crude approximation of the quantity of coagulant required for charge neutralisation can be gained from a simple charge balance (see Section 1.5.5). A more sophisticated model is described in Dentel (1988). Although the latter was developed for alum, it appears to have wider application (Dentel, 1991). When using hydrolysing coagulants (Section 1.5), the core difficulty lies in the identification of the coagulant product responsible for charge neutralisation. The species concerned can either arise from the dissolved phase (*e.g.* in the form of a monomer or oligomer) or may arise from the solid phase prior to adsorption. Nucleation and hydrolysis can also take place on the surface. These features are illustrated in Figure 1.12. Although attention has focussed on the role of dose in charge neutralisation, it must be emphasised that pH is equally significant, through its impact on the ionisation of surface functional groups and on the solubility chemistry.

A fuller account of the status of Figure 1.11 in the context of destabilisation using aluminium(III) and iron(III) salts is given in Stumm and O'Melia (1968) and O'Melia (1972).

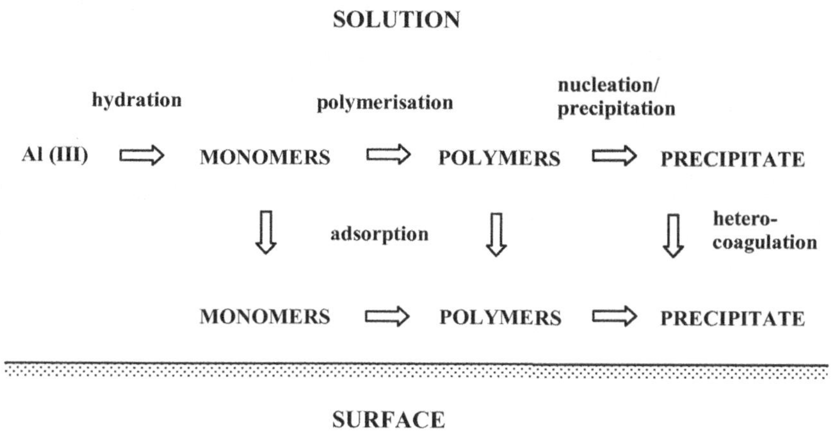

Figure 1.12. Schematic representation of pathways associated with charge neutralisation (Dentel, 1991).

1.3.2.3 Enmeshment

The widespread view of enmeshment (also known as sweep coagulation) is one in which the raw water colloids become enmeshed in a voluminous precipitate gained from aluminium and iron coagulants. It is a widely used method for destabilising turbid suspensions. Beyond the basic chemistry, its mode of action has been barely probed.

When coagulating turbid suspensions, the doses employed are generally well in excess of those associated with charge neutralisation and above the solubility limit of the products in use (see Figure 1.22). The mass of precipitate liberated *e.g.* Figure 1.33 depends primarily on dose and pH, but is also influenced by the anion content of the coagulant. The bulk precipitate is a heterogeneous mix of trihydroxide and small polycations carrying positive charges. Collectively the precipitate is often referred to as $Al(OH)_3(s)$ or $Al(OH)_3(am)$ and similarly for iron. The symbols (s) and (am) refer to the solids phase and amorphous state respectively. These conventions (which will be used elsewhere in the text) disguise both the heterogeneous nature of the precipitate and its charge (see Figure 1.31 and Table 1.4). The precipitate is naturally flocculant, a feature that enables it to form clusters and simultaneously gathering raw water primary particles. Sweep floc is particularly advantageous in the coagulation of low turbidity waters, because the precipitates add to the particle numbers, this boosting the flocculation rate (Section 1.3.3).

Figure 1.13. Coatings at 3 mg Al / l (Bache et al., 1999).

Figure 1.14. Coatings at 9 mg Al / l (Bache et al., 1999).

By the time the primary particles have been gathered into floc structures, their surfaces will have been covered by a strongly adherent positively charged precipitate. In spite of significant zeta potential (*e.g.* Figure 1.8 at 5 mg/l), the primary particles bond well and form coagulation structures. Bache *et al.* (1999) attributed this phenomena to electrostatic bridging (Section 1.2.9) - the crucial evidence being the association between floc strength and the primary charges. At low coagulant dose (Figure 1.13) the particles are covered by a glistening coating. Close inspection shows evidence of small linkages between particles. At higher dose, the artefacts shown in Figure 1.14 stem from dried bulk precipitate. This has an entirely different appearance and character (mobility) when compared with the coating shown in Figure 1.13. The bulk precipitate forms a space-filling gel (Figure 1.9) and only plays a minor role in the control of floc strength. Nevertheless, overdosing leads to reductions in floc strength - believed to be caused by excess precipitate interfering with electrostatic bridging (Bache *et al.*, 1999). The gel has a low yield stress, but is protected from external shear by the stronger structure formed by the primary particles. The gel appears to play a prominent role in dewatering (Section 3.3.7). Flocs formed by enmeshment tend to be large, facilitating settlement.

Overall, one sees that enmeshment has many facets which tend to be overlooked. The mechanisms described warrant greater scrutiny.

1.3.2.4 Polymers and destabilisation

Natural and synthetic polymers are used both for destabilisation purposes and as a floc aid, the latter taking advantage of its bridging capacity (see also Sections 3.2.2.6 and 3.2.6.1) The behaviour of polymers is dependent on a multitude of factors including polymer type, molecular weight, charge density, ionic strength as well as features of the receiving surface (Section 1.6). Generally there is no single mechanism of destabilisation. Rather there is a combination of mechanisms in which one may dominate. In all cases, the polymer needs to be adsorbed - adsorption being governed by both electrostatic and non-electrostatic factors.

When the polymer and colloid are oppositely charged, increasing dosage is accompanied by charge reduction, this helping to destabilise the suspension. Excess doses may result in charge reversal and restabilisation. This is partly due to increases in the energy barrier, but may also be caused by steric effects (Section 1.2.7). Polylectrolytes tend to have a flat conformation due to mutual repulsion between charged segments, and will tend to adsorb in the form of electrostatic patches as illustrated in Figure 1.7.

Where the polymer and particles have similar charge, or cases in which a non-ionic polymer is used, bridging is likely to be the dominant mode of destablisation. Two conditions are necessary for bridging. First, the polymer needs to be of sufficient length to span the gap associated with the energy barrier. Second, there must be available sites on the adjacent particle to anchor other segments of a polymer chain. Healy and La Mer (1964) proposed that the optimum condition for this form of bridging would occur when the particle surfaces were half-covered with adsorbed polymer, thereby maximising the opportunity for similar particle-polymer combinations to become attached to the vacant sites.

1.3.3 Principles of flocculation

1.3.3.1 Introduction

The rate of aggregate growth depends on both the inter-particle collision frequency and on whether collisions result in adhesion. For binary collisions, the collision frequency per unit volume (J_{ij}) between n_i particles per unit volume of size i and n_j particles per unit volume of size j is given by:

$$J_{ij} = k_{ij} n_i n_j \tag{1.10}$$

in which k_{ij} is a rate constant. An important feature of this relationship is the sensitivity of the contact rate to the numbers of particle present - high contact rates being favoured by high number concentrations. The fraction of collisions leading to successful aggregation is represented by the inclusion of a *collision efficiency factor* (α) or *trapping efficiency.* If conditions are very unfavourable, *e.g.* arising from the presence of stong repulsive forces, $\alpha \approx 0$, whereas, if there is no energy barrier and attractive forces dominate then $\alpha \rightarrow 1$. In the latter circumstances, the rate of aggregate formation is at its most rapid and controlled by transport factors.

When particles are small, the transport rate (J_{ij}) is driven by Brownian motion *i.e.* particle movements induced by molecular bombardment. Associated aggregation is termed *perikinetic aggregation*. Where collisions are brought about by bulk fluid motion, the term *orthokinetic aggregation* applies. If sedimentation dominates, collisions and aggregation arising from particles settling at different rates is called *differential settlement*. Perikinetic aggregation applies to particles smaller than about 1 μm. For larger particles, and in the absence of quiescent conditions favouring sedimentation, orthokinetic processes dominate.

The rate of aggregation determines the characteristic timescale of the process. In practice, flocculation times are generally less than 30 min.

1.3.3.2 Perikinetic aggregation

The classical framework for modelling perikinetic aggregation stems from Von Smoluchowski (1917). Since then, the approach has been much elaborated to take account of colloidal and hydrodynamic interactions, the fractal nature of flocs as well as other factors. Here, the review is restricted to foundation concepts with the intention of identifying controlling parameters and timescales. For a fuller account, readers are referred to Sonntag (1993).

The simplest scenario is the case of a collection of particles of the same size at a concentration of n particles per unit volume. When a collision occurs, the probability of doublet formation is represented by the collision efficiency factor. A consequence of aggregation is a reduction in the primary particle concentration, this being expressed by:

$$\frac{dn}{dt} = -\alpha J_{ii} = -\alpha k_p n^2 \tag{1.11}$$

Because particles are identical, they have the same rate constant *i.e.*

$$k_p = \frac{4kT}{3\mu} \tag{1.12}$$

in which k is the Boltzmann constant, T the absolute temperature and μ the dynamic viscosity of the medium. Subject to the initial condition $n = n_o$ at $t = 0$, integration of Equation (1.11) leads to:

$$n = \frac{n_o}{1 + \alpha k_p n_o t} \tag{1.13}$$

A representative timescale, τ (the half-life) is given by the time at which the initial particle concentration, n_o, is reduced by one half, this being:

$$\tau = \frac{1}{\alpha k_p n_o} \tag{1.14}$$

Although this particular model is grossly simplified and ignores the evolution of triplets *etc*, it illustrates the critical dependencies that are encountered in more complex scenarios. For water at 20°C, k_p is on the order of 5×10^{-18} m^3 s^{-1}. For the case $\alpha = 1$ and $n_o \sim 10^{11}$ cm^{-3}, Equation (1.14) shows $\tau \sim 1$ s. Particle concentrations in treated waters are generally much less than the cited value, though this can be exceeded in the early stages of sweep coagulation when using an Al(III) salt. When the trapping efficiency is low ($\alpha \rightarrow 0$), this has a profound influence on the coagulation timescale.

The collection efficiency is influenced by colloidal and hydrodynamic interactions. Of these, the colloidal interactions have the greatest significance. Following work of Fuchs (1934), the influence of an energy barrier on the aggregation rate is represented by a stability ratio (W) defined by:

$$W = \frac{1}{\alpha} = \frac{\text{Number of collisions between particles}}{\text{Number of collisions that result in aggregation}} \tag{1.15}$$

In the absence of an energy barrier, the rate of aggregation can be represented by the term R_f, the subscript referring to *fast aggregation i.e.* limited only by the rate of diffusion of the particles towards one another. In essence this is set by Equation (1.11) for the case $\alpha = 1$. When an energy barrier is present, the rate of slow aggregation (R_s) is defined by:

$$R_s = R_f / W \tag{1.16}$$

For spherical particles of different radii a_i and a_j and subject to double layer repulsion and van der Waals attraction, Reerink and Overbeek (1954) showed that:

$$W \approx \frac{1}{\kappa(a_i + a_j)} \exp\left(\frac{V_{max}}{kT}\right) \tag{1.17}$$

where κ is the Debye-Huckle parameter (Equation 1.4) and V_{max} is the height of the energy barrier. For modest barrier heights of just a few kT, Equation (1.17) shows that W can take on very high values. An example is cited in Elimelech *et al.* (1995, p172) showing the effect of the barrier on 1 μm particles in a 100 mM solution of a 1:1 electrolyte, for which the term κa is about 500. When V_{max} = 10kT, W ≈ 45 whereas when V_{max} = 20kT, W ~ 10^6, implying that α ~ 10^{-6}. In physical terms, α represents the fraction of the collisions with kinetic energy sufficient to exceed the potential energy barrier.

The hydrodynamic factor mentioned above arises from fluid displacement and drainage from the gap between two particles as they approach each other. This tends to slow the aggregation rate. The precise value of W depends on the Hamaker constant (see Elimelech *et al.*, 1995). Generally W ≈ 2, indicating a halving of R_f, or a doubling of the aggregation timescale; this is more or less in accord with experimental evidence.

1.3.3.3 Orthokinetic aggregation

In orthokinetic aggregation, the relative particle motion which results in a flux of particles towards another is determined by the *local* velocity gradient or strain rate $(\varepsilon/\nu)^{1/2}$, in which ε is the rate of energy dissipation per unit mass and ν the kinematic viscosity. In stirred reactors, this factor is often represented by the 'mean velocity gradient' (G) (Equation 1.39), but the former definition is more appropriate. Once again, the rate of aggregation stems from Equation (1.10). For illustrative purposes, we adopt the same approach as employed in Section 1.3.3.2 by focussing on a regime of identical particles of diameter, d, and number concentration, n. For this scenario, the Von Smoluchowski (1917) approach may be used with a rate constant specified by:

$$k_o = \frac{2}{3} G d^3 \tag{1.18}$$

From Equation (1.10) and again introducing a collision efficiency factor (α), the aggregation rate is specified by:

$$\frac{dn}{dt} = -\frac{2}{3} \alpha G d^3 n^2 \tag{1.19}$$

The volume fraction within the suspension can be expressed as:

$$\phi = \frac{\pi}{6} d^3 n \tag{1.20}$$

If this remains constant during flocculation, it allows Equation (1.19) to be treated as pseudo-first order reaction of form:

$$\frac{dn}{dt} = -\frac{4}{\pi}\alpha\,\phi G n \tag{1.21}$$

By applying the boundary condition n = n_o at t = 0, integration of Equation (1.21) yields:

$$n = n_o \exp\left(-\frac{4}{\pi}\alpha\phi G t\right) \tag{1.22}$$

A representative timescale (τ) can be defined by the condition n = $n_o e^{-1}$ *i.e.*

$$\tau = \frac{4}{\pi}.\frac{1}{\alpha\phi G} \tag{1.23}$$

The salient feature of Equation (1.23) is the inverse dependence between the aggregation timescale and the parameters α, ϕ and G. In this description the major unknown is α. Before discussing its status, it is worth noting that when α is constant, the product $\phi G\tau$ is also constant. Thus for a fixed flocculation time (*e.g.* corresponding to retention time of a flocculator in steady flow conditions), it appears that changes in ϕ can be compensated by adjustments in G.

With regard to the collision efficiency, it must be recognised that when particles approach one another, they are subject to force fields of the type described in Section 1.2.1, as well as those arising from the interaction of the flow fields *surrounding* the moving particles. For example, consider a small particle approaching a larger particle within the mutual collision cross-section. In the absence of external forces, a linear trajectory would be maintained and the particles would collide. If aggregation were successful then α = 1. However, with force fields present, the small particle is more likely to follow the streamlines round the larger particle and the collision probability will be reduced.

The upshot is that the collision efficiency depends on the relative size of the interacting particles and the prevailing force fields. Where particles have been destabilised, α is at its largest for particles of similar size, and may become exceedingly low *e.g.* $\alpha < 10^{-4}$ where there are large differences in size. Further detail is provided in Adler (1981) and Han & Lawler (1992).

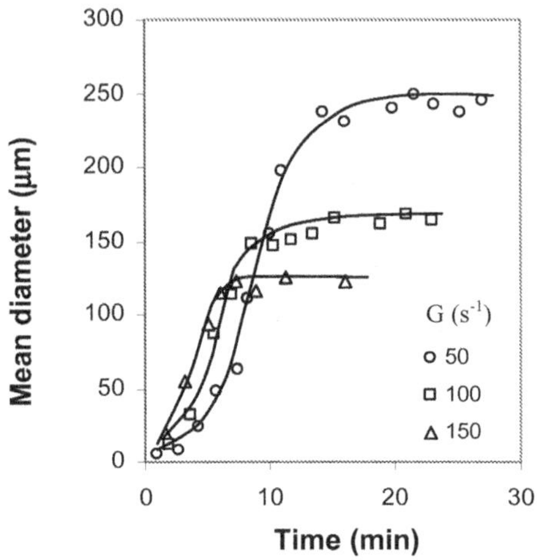

Figure 1.15. Floc growth and velocity gradient (adapted from Flesch *et al.* 1999).

Data in Figure 1.15 illustrates the response of particle growth to G. In the early stages, it is seen that the most rapid growth occurs at the highest G value, this being in accord with the kinetics described by Equation (1.19). However at later times, it is seen that the growth appears to cease with the existence of a limiting size whose magnitude depends on G. The reason that growth appears to cease is because aggregates are being broken as fast as they form. Schematically, this can be explained by including a further term into the kinetics process to allow for breakage *i.e.*

$$\frac{dn}{dt} = -\alpha k_o n^2 + k_R n \tag{1.24}$$

in which k_R is a breakage (rupture) rate constant. When growth and rupture are in balance, dn/dt = 0. There are differing views as to how to represent k_R, some being addressed in Kramer and Clark (1999). The analysis of Vivaldo-Lima *et al.* (1998) also points to the significance of non-homogeneities in the turbulence regime.

For continuous flow systems as found at full scale, the growth kinetics need to be examined in a slightly different way because they are in quasi-steady state. The scenario that is faced is one in which microflocs produced by the combination of rapid mixing and perikinetic aggregation come into contact with a distribution of large flocs held in a contact process such as a flocculator or floc blanket clarifier. This was examined in Tambo and Hozumi (1979). Their logic is similar to that described by Equation (1.22), but with ϕ ($\equiv \phi_F$) representing the volume fraction of the large flocs and n referring to the microflocs.

The task of accounting for the many factors involved is complex and can only be approached by use of numerical models, such as Vivaldo-Lima *et al.* (1998). In contrast to the knowledge and understanding of the rate constant k_{ii} and α, significant problems remain with the representation of rupture processes, the properties of the resulting fragments and the ways in which they re-enter the growth cycle.

1.4 MIXING FACTORS

1.4.1 Introduction

Coagulation and flocculation pose different demands on mixers. Destabilisation requires effective dispersion so that the coagulant or coagulant products reach all particles in the suspension. Poor mixing can lead to the inefficient use of chemicals, with some portions of the suspension being under-dosed and others overdosed, or perhaps arriving in an inappropriate form. Commonly used coagulants such as Al(III) and Fe(III) salts, undergo rapid hydrolysis, their charge status changing with time. If the quest is to take advantage of particular species for destabilisation, there may be a race between the reaction kinetics and the mixing timescales. When more stable compounds, *e.g.* polyaluminium chloride (PACl) or organic polymers, are used, the principal problem is adequate dispersion. Rapid mixing has the twin benefit of enhancing dispersion and accelerating the initial stages of growth. Thereafter, flocculation requires a gentler form of mixing. In all practical cases, the flow will be fully turbulent.

For dispersion, the key task is to deliver chemicals to point of use in all places. The ideal is to have complete uniformity *i.e.* applying to molecular scales. This involves bulk mixing at large scales, turbulence delivering materials to progressively smaller scales - the final phase being controlled by molecular diffusion (Figure 1.16). In turbulent motion, this transfer mechanism reflects the way in which turbulence energy is transferred from large scale fluid motions to the smallest scales, where it is dissipated as heat (Figure 1.17). Figure 1.16 indicates some of the critical length scales which feature in mass transport and the description of turbulence. Perhaps the most important parameter implicit in this framework, is the rate of energy dissipation per unit mass (ε). This underpins many issues discussed in the text and will be discussed below before providing further detail of mixing.

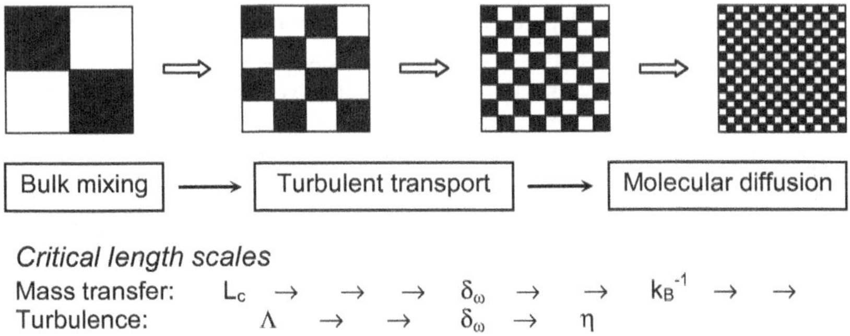

Figure 1.16. Schematic representation of transfer of a component (*e.g.* blackened areas) from large scales to small scales resulting in a more even distribution. Also shown are length scales associated with energy and mass transfer by turbulence (diminishing left to right) - the destruction of concentration segregation being tied to the hierarchy of eddy sizes and straining motions. Symbols are introduced and specified in Section 1.4.2.

1.4.2 Facets of turbulence

1.4.2.1 Energy cascade

Turbulence is often envisaged as a hierarchy of irregular motions containing vorticity on different scales or eddy size. Fluid elements are subject to continuous deformation, whereby any small spherical element is ultimately drawn to a thin ribbon. Such deformations cause vortex stretching, a mechanism that leads to the transfer of turbulent kinetic energy from the larger eddies to smaller eddies. The largest eddies, say of size Λ, gain energy from the mean motion. These degenerate to form a larger group of smaller eddies and so on, as illustrated in Figure 1.17.

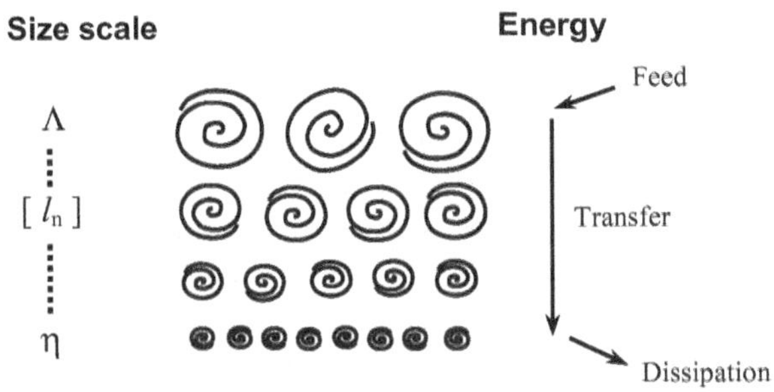

Figure 1.17. Schematic view of energy routing (based on Herweijer, 1995).

Provided eddies are sufficiently large, the energy transferred from one generation to the next, far exceeds any losses associated with viscous dissipation. If we use the subscript n to identify the nth generation, the rate of energy transfer from the nth to the (n+1)[th] generation can be represented by:

$$\varepsilon_n \sim \frac{u_n^2}{t_n} \tag{1.25}$$

in which u_n^2 scales the kinetic energy and t_n is a time scale defined by:

$$t_n \sim \frac{l_n}{u_n} \tag{1.26}$$

The timescale (*time of turnover*) indicates the time during which the eddies lose a significant fraction of their kinetic energy. Equations (1.25) and (1.26) combine to yield:

$$\varepsilon_n \sim \frac{u_n^3}{l_n} \tag{1.27}$$

For stationary turbulence, the energy transfer rate applies to all scales and is equal to the mean dissipation rate $<\varepsilon>$ *i.e.*

$$\varepsilon_n = <\varepsilon> \tag{1.28}$$

Here the brackets < > refer to a time average value. An immediate consequence of Equation (1.28) is that the velocity scale u_n associated with the size scale l_n, is defined by:

$$u_n \sim (<\varepsilon> l_n)^{1/3} \tag{1.29}$$

For smaller eddies in which viscous dissipation is significant, the time scale expressed by Equation (1.26) is no longer valid. Instead, it is defined by the ratio λ/u in which λ is the Taylor length scale. For isotropic turbulence, λ is defined by the relationship:

$$\varepsilon = 15\nu \frac{u^2}{\lambda^2} \tag{1.30}$$

showing that the timescale is scaled by the factor $(\nu/\varepsilon)^{1/2}$. The smallest length scale of turbulence is represented by the Kolmogorov length defined by:

$$<\eta>=\left(\frac{\nu^3}{<\varepsilon>}\right)^{1/4} \tag{1.31}$$

A common interpretation of η is that it represents the smallest eddy size. While this may be roughly true, its size reflects the condition in which inertial forces and viscous forces are of similar magnitude. This is shown by the Reynolds number:

$$\mathrm{Re}=\frac{\mathrm{v}_K\eta}{\nu}=1 \tag{1.32}$$

in which v_K is the Kolmogorov velocity scale defined by:

$$\mathrm{v}_K=(<\varepsilon>\nu)^{1/4} \tag{1.33}$$

The associated time scale for the dissipative structure of turbulence is:

$$\tau_K=(\nu/<\varepsilon>)^{1/2} \tag{1.34}$$

Equation (1.34) applies to all length scales smaller than the Kolmogorov length. Fluid accelerations and the associated dynamic forces are at their most intense in the domain $d \sim <\eta>$ (Levich, 1962).

1.4.2.2 Smallest stable vortex

Although the Kolmogorov length is indicative of the scale at which the viscous dissipation of turbulent energy is significant, it must be recognised that viscous dissipation coexists with vortex structures on a larger scale. Tennekes and Lumley (1994; p 72) argued that the length scale (δ_ω) of the associated vortex motion must be proportional to $<\eta>$. Baldyga and Bourne (1984) estimated the value as $\delta_\omega = 12 <\eta>$. The significance of this particular size (or thereabout) is that the vorticity (or rotation rate) of eddies larger than $12 <\eta>$ is increased by the straining action and energy transfer from still larger eddies, whereas the vorticity of eddies smaller than $12 <\eta>$ is progressively reduced by viscosity and eddies will cease to exist. Baldyga and Bourne (1984) suggested that that eddies of 12 to 15 $<\eta>$ may be regarded as the smallest stable vortex size and forms an integral part of the fine dissipating structure of turbulence. They argued that this class of eddies plays a critical role in micromixing. This is promoted by the process of engulfment, whereby fluid gets sucked into the vortices as illustrated in Figure 1.18.

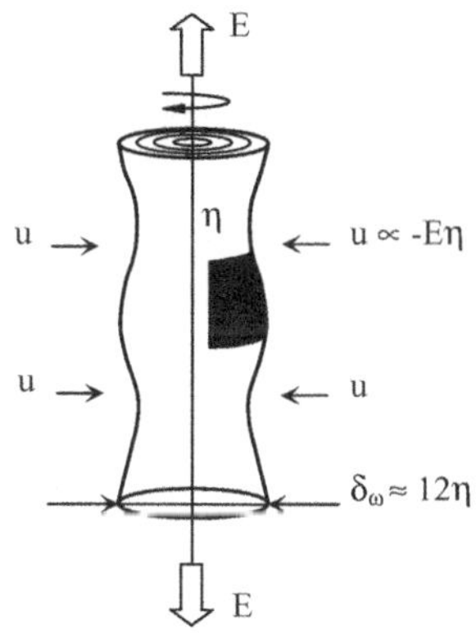

Figure 1.18. Vortex stretching causes the engulfment of fluid and suspended materials from the surroundings. E refers to the strain rate and δ_ω to the size of the smallest stable vortex (Baldyga and Bourne, 1995).

A consequence of this process (detailed in Baldyga and Bourne, 1984) is the formation of partially segregated laminae which are held within the vortices. During vortex stretching, these ribbon-like structures become thinner, making them more amenable to segregation and reduction by molecular diffusion. This ends as a uniform concentration at the molecular scale. During engulfment, it is inevitable that flocs (often smaller than 12 $\langle\eta\rangle$) get transported into the vortices. Once there, they are subject to stretching at a rate proportional to $(\langle\varepsilon\rangle/\nu)^{1/2}$ and under the influence of the kinetic energy within the spinning vortex. These conditions are conducive to rupture.

1.4.2.3 Energy spectrum

Equation (1.29) indicates the relationship between an energy component $u(l)^2$ and its size scale. In a stationary turbulence field all size scales coexist and contribute to the total turbulence kinetic energy per unit mass at a particular location. For isotropic turbulence this is given by:

$$\frac{3}{2}u^2 = \int_0^\infty E(k,t)\,dk \tag{1.35}$$

in which E(k,t) is a power spectrum with k as the wave number.. Tennekes and Lumley (1994) considered that an eddy of wave number k could be envisaged as a form of disturbance containing a spike of energy (~kE) which contributes to the power spectrum. From this it was argued $l \sim k^{-1}$. In the context of Equation (1.29), this leads to:

$$kE \sim \langle u(l)^2 \rangle \sim \langle \varepsilon \rangle^{2/3} k^{-2/3} \tag{1.36}$$

showing:

$$E \sim \langle \varepsilon \rangle^{2/3} k^{-5/3} \tag{1.37}$$

This is the famous Kolmogorov 5/3 law. Although its form has been indicated for eddies complying with Equation (1.29) (the inertial range), the 5/3 form holds over a larger range of wave numbers.

At wave numbers greater than k_η ($= \eta^{-1}$), $E(k,t) \propto k^{-7}$ indicating that turbulence kinetic energy rapidly diminishes at very small length scales because of the effects of viscosity. In Figure 1.19, k_o corresponds to the wave number at which E(k,t) is at its maximum value (see Hinze, 1987). When $k < k_o$, $E(k,t) \rightarrow 0$.

1.4.2.4 Decay of turbulence

From the above, it is apparent that ε plays a central role in the description of turbulence and its properties. Amongst the energy-containing eddies, forming the peak of the power spectrum (Hinze, 1987):

$$\frac{\varepsilon t^2}{\nu} = \text{constant} \tag{1.38}$$

The constant is a form of Reynolds number which remains constant during the decay of turbulence (see also Bache and Rasool, 2001). Equation (1.38) shows $\varepsilon \propto t^{-2}$, implying that turbulence decreases with time. In consequence it must be recognised that ε decays rapidly through space - this affecting the many parameters which depend on ε.

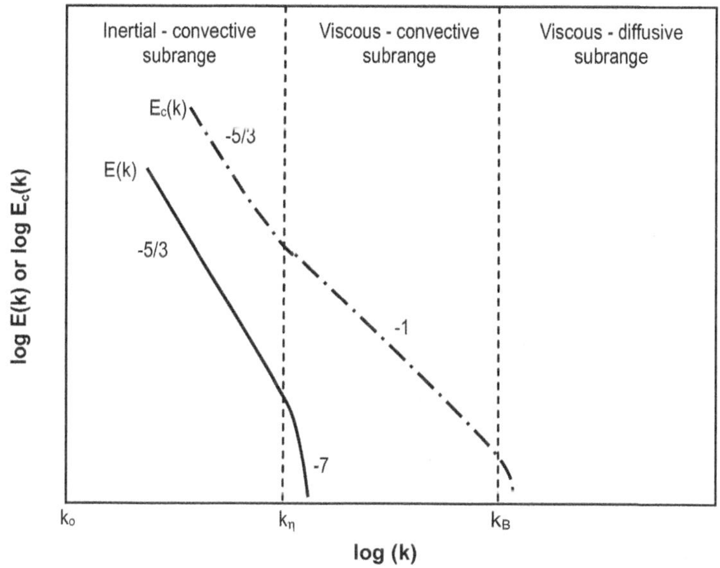

Figure 1.19. Velocity and concentration spectra (based on Brodkey, 1967).

1.4.3 Characterisation of mixers

Power input

In turbulent flows, the rate of energy dissipation per unit mass emerges as a critical parameter for both expressing and scaling character of the turbulence. In a vessel of fluid volume V, with ρ as the fluid density, the spatial mean value of ε and the 'mean velocity gradient' G are controlled by the power input (P) per unit mass into the fluid *i.e.*

$$\bar{\varepsilon} = \frac{P}{\rho V} = \nu G^2 \tag{1.39}$$

In pipes and channel flows the power input is expressed by:

$$P = \rho Q g h \tag{1.40}$$

where Q is a flow rate, g the gravitational acceleration and h the associated head loss. According to Edwards (1985), Equation (1.39) can also be applied to mixing vessels stirred by an impeller because the impeller generates a circulation flow which is responsible for bulk transport throughout the mixing vessel. Most of the energy dissipation takes place in the vicinity of the impeller where ($<\varepsilon>/\bar{\varepsilon}>1$) (*e.g.* Figure 3.9) and decays in regions away from the impeller where ($<\varepsilon>/\bar{\varepsilon}<1$). Mixing is at its most intense where $<\varepsilon>$ is greatest. Thus when dispersion is the objective, it is highly desirable to circulate the liquid as frequently as possible through regions of high intensity turbulence, this being a benefit of the circulation flow.

1.4.4 Mixing timescales

1.4.4.1 Bulk mixing

If a pulse of a tracer is injected into a stirred tank, then the measured concentration (C) at a sample position is likely to follow the pattern illustrated in Figure 1.20(b), this resembling a damped oscillation. The periodicity t_c corresponds to the circulation or turnover time of the bulk fluid. Through time, the concentration converges to an average value $<C>$. The instantaneous concentration can be represented by $C(x,t) = <C> + c(x, t)$ in which $c(x, t)$ is a fluctuating component with root mean square value c'. The variance c'^2 describes the segregation of a concentration field and the quest is to identify the associated time scale.

Generally it is found that the mixing time t_M corresponds to a few circulations, by which time the variations have become strongly damped *e.g.* Godfrey and Amirtharajah (1991). Dimensional analysis indicates that Nt_c or Nt_M become constant at high Reynolds Number, the constant depending on geometrical factors and N referring to the rotation rate.

For an impeller of diameter D_m, the power input complies with the scaling:

$$P = P_o \rho N^3 D_m^5 \tag{1.41}$$

in which P_o is the power number (tends to constant value in fully turbulent flow). By substituting Equation (1.41) into Equation (1.39), it may be shown that $G \propto N^{3/2}$. Since $t_c \propto N^{-1}$, it follows that turnover or mixing time complies with the dependence:

$$t_c \propto G^{-2/3} \tag{1.42}$$

1.4.4.2 Dispersive and diffusive mixing

This phase of mixing is strongly associated with the decay of turbulence and the effects of shear. Although eddy transport effectively ceases at the Kolmogorov length, straining motions continues down to the molecular scale at which molecular diffusion becomes dominant. The scale at which convection due to straining action and molecular diffusion become of the same order of magnitude is indicated by the Batchelor wave number $k_B = (<\varepsilon>/\nu D_c^2)^{1/4}$ in which D_c is the molecular diffusion coefficient of the tracer. The domain of wave numbers between k_η and k_B is known as the viscous-convective sub-range. Table 1.3 shows values of η and k_D^{-1} for the illustrative values $<\varepsilon> = \bar{\varepsilon}$ and $D_c = 10^{-3}$ m^2 s^{-1}. Particular note should be taken of the interval between η and k_B^{-1}, emphasising the importance of straining motion supporting mass transfer in the viscous-convective range.

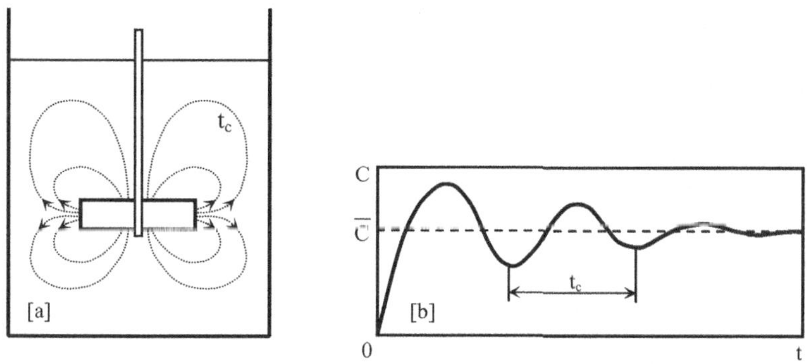

Figure 1.20. (a) Illustrating flow patterns and circulation in a stirred tank and (b) the tracer concentration as a function of time in a batch operated tank.

Table 1.3. Illustrating sensitivity of η and k_B^{-1} to magnitude of G.

G (s^{-1})	10	100	1000
$\bar{\varepsilon}$ (m^2 s^{-3})	10^{-4}	10^{-2}	1
η (μm)	316	100	3.2
k_B^{-1} (μm)	10	3.2	1

When the Schmidt Number Sc (= ν/D_c) is high, the power spectrum follows the dependence $E_c(k) \propto \varepsilon_c k^{-1}$ in which ε_c to the rate of destruction of the c'^2, this being analogous to ε in the destruction of u'^2. Corrsin (1964) analysed the decrease of segregation in decaying turbulence. The analysis was based on the similarity between the velocity field and the scalar field in conjunction with the equation:

$$\frac{dc'^2}{dt} = -\frac{c'^2}{\tau_c} \tag{1.43}$$

When τ_c is independent of time, Equation (1.43) leads to $c'^2 = c_o'^2 \exp(-t/\tau_c)$ in which $c_o'^2$ is an initial segregation and τ_c the time constant. Corrsin's analysis led to the deduction:

$$\tau_c = \frac{1}{2}\left[4\left(\frac{L_c^2}{\varepsilon}\right)^{1/3} + (\nu/\varepsilon)^{1/2}\ln Sc\right] \tag{1.44}$$

where L_c is an integral length scale for the concentration field. In Equation (1.44) the first term on the right hand side gives the timescale for the destruction of segregation through the inertial–convective range (Figure 1.19), whereas the second term corresponds to the timescale associated with the destruction of segregation in the viscous-convective size range. For dilute aqueous suspensions $Sc \sim 10^3$ and the timescale is dominated by the first term *i.e.* $\tau_c \approx 2(L_c^2/\varepsilon)^{1/3}$, this showing that τ_c behaves in accord with the dependence:

$$\tau_c \propto G^{-2/3} \tag{1.45}$$

1.4.4.3 Implications

Figure 1.21 compares typical mixing times with some metal ion hydrolysis timescales in a hypothetical 100 litre Rushton reactor. Mixing timescales were estimated using the principles described above. In stirred reactors, the turnover time is well in excess of the reaction timescales implying that they will be very inefficient in situations where the reaction timescale is faster. It is here that in-line blenders are superior. The Corrsin timescale will apply to all mixer types. When G values are comparable, differences will depend on the value of the integral scale (L_c). This is likely to be smaller in a more compact mixer - again favouring in-line blenders. For the situation described in Figure 1.21, the Corrsin timescale is about 10x faster than the bulk turnover time, but still falls short of the cited reaction timescales. This can have a crucial bearing on the mode of destabilisation when using Fe(III) and Al(III) salts, a subject which is pursued further in Section 1.5. Rapid mixing is discussed in Section 3.2.7.1.

1.5 HYDROLYSING COAGULANTS

1.5.1 Perspective

The character of a floc is moulded by the surface chemistry, structure and mechanical properties of the materials from which it is formed. For sludges derived from the coagulation of low-turbidity coloured waters using alum, Warden (1984) reports that about 45% of the dry mass arises from the coagulant and about 11% for turbid waters. Such data emphasises that coagulant products can represent a substantial fraction of the floc solids. There exist a vast array of coagulants and the blunt truth is that we know little about their form and status beyond the coagulation phase.

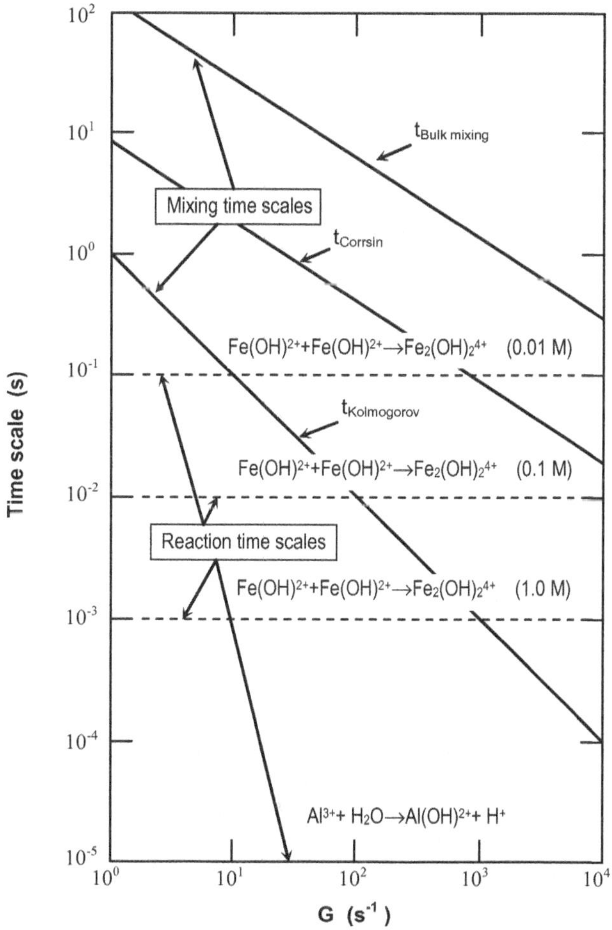

Figure 1.21. Comparison of mixing and reactive time scales in 100 litre Rushton backmix reactor (adapted from David and Clark, 1991). Timescales are defined by Equations (1.42), (1.45) and via (1.34).

Here we are mainly concerned with the nature of precipitates arising from the commonly used forms of aluminium and iron. A major feature of the hydrolysing metal coagulants is that they react with water at a far greater rate than with contaminants - resulting in an array of metal hydroxides. Coagulation depends on the impact of both soluble and colloidal forms and there is often great uncertainty as to which of the many species in existence is responsible for a particular outcome.

The solids phase is a heterogeneous mix of compounds and collectively referred to as the bulk precipitate. Besides playing a role in coagulation, it has an innate tendency to form structures. This particular feature also applies to larger polycations within the dissolved phase. Both are a direct consequence of hydrolysis, in which there is the progressive reduction in the effective charge per metal ion as hydrolysis proceeds. Through time, this permits hydrogen bonding and van der Waals interactions to gain greater prominence, permitting the formation of a *sol* (small clusters). These in turn can aggregate further to form larger units and conditions permitting, progress towards the infinite phase (a *gel*). Structures of this type control strength and many properties of flocs gained from the coagulation of low turbidity coloured waters. In cases of flocs formed from turbid suspensions using the hydrolysing coagulants, the precipitate exists as a space filling gel as illustrated in Figure 1.9. The structure of the material filling the interstices affects settleability and dewaterability through its impact on water content and permeability (Section 3.3.7). Another important attribute of precipitates is their adsorptive capacity, this being linked to their charge status and surface area.

1.5.2 Monomeric hydrolysis products

Monomers tend to dominate the solubility chemistry at the concentrations of Al and Fe in common use. In concentrated solution at low pH, alum and ferric salts exist in the hydrated form $[Me(H_2O)_6]^{3+}$, this referring to the cation and its primary hydration shell. The water molecules are strongly polarised and held in an octohedral coordination. With increasing pH, there is a step-wise substitution of the coordinated water by other ligands. The complexes that tend to dominate are those comprising H_2O and OH^- ligands. The changes of H_2O to OH^- within the coordination structure can be brought about either by the dissociation of the H_2O ligand or from the replacement of H_2O by OH^- ligands. The extent to which OH^- ions become bound, is dependent on the concentration of OH^- ions in solution, and thereby on pH. The first step in the hydrolysis process is shown by:

$$\{Me(H_2O)_6\}^{3+} \leftrightarrow \{Me(H_2O)_5OH\}^{2+} + H^+ \tag{1.46}$$

Omitting the coordinating water molecules, this leads to the hydrolytic sequence:

$$Me^{3+} \rightarrow Me(OH)^{2+} \rightarrow Me(OH)_2^{+} \rightarrow Me(OH)_3 \rightarrow Me(OH)_4^{-} \quad (1.47)$$

in which it is seen that the charge is progressively reduced and ultimately reversed. When the concentration of the solution exceeds the solubility limit of the metal hydroxides, they coexist as a group in equilibrium with amorphous hydroxide precipitate. Their molar concentrations depend on a range of factors including pH, ionic strength, temperature and the solubility of the precipitate. The latter is often system specific.

Figure 1.22 is based on data cited in Duan and Gregory (2003) and shows the concentrations of the monomers and the trends for the total presence (based on the sum of the monomeric concentration). These give an indication of the expected metal solubility. In each case it is seen that there is a minimum solubility at a certain pH value. For aluminium, the minimum solubility is about 1 µM at a pH of about 6. For iron, the minimum solubility is much lower and occurs at higher pH. Generally the aluminium solubility is more sensitive to pH. At very low and very high pH values, the solubility is determined by the behaviour of Me^{3+} and $Me(OH)_4^{-}$ respectively. The plots do not include influence of polymers or the effect of anions on the total solubility. The operating regions occur in the over-saturation domain, indicating that precipitates are always likely to present.

Figure 1.23 shows the fractions of dissolved monomers of Al and Fe as a function of pH. The relative abundance of the species depends on the prevailing equilibrium constants, the data being illustrative. In the case of Al, it is seen that the $Al(OH)^{2+}$ and $Al(OH)_2^{+}$ tend to be the dominant species in the pH range 5 - 6.5. In the case of Fe, it is seen that $Fe(OH)_2^{+}$ dominate in the pH range 3.5 - 6.0. Echoing remarks made in Edzwald and Tobiason (1998), it is emphasised that the distribution of charged Al and Fe monomers have a profound effect on the coagulation of negatively charged particles and macromolecules such as aquatic humic matter. It must be stressed that the cited distributions are sensitive to temperature and ionic strength. Further, the presence of other ionic species, notably anions, may influence the distributions shown in Figure 1.23 and hence the pH for minimum solubility.

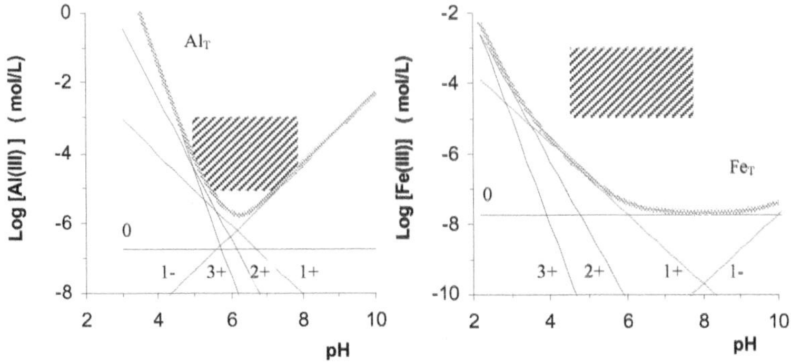

Figure 1.22. Concentrations of monomers in equilibrium with isomorphous precipitates of Al(III) and Fe(III) at zero ionic strength and 25 °C. Labels indicate total presence (Me_T) and monomer type (shown by species charge). Shaded areas are approximate operating regions in water treatment practice (adapted from O'Melia, 1972).

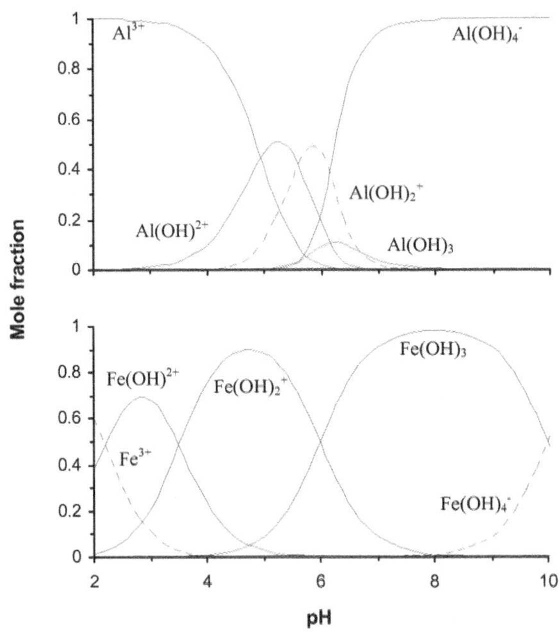

Figure 1.23. Fractions of dissolved Al and Fe monomers in equilibrium with amorphous hydroxides based on data in Figure 1.22 (based on Duan and Gregory, 2003).

1.5.3 Polynuclear species

1.5.3.1 Hydroxide bridges

Polynuclear species represent an important stepping stone in the transition from the monomeric forms to the formation of the precipitate. Their presence is favoured by higher concentrations of the metal ion, longer aging times or elevated temperature. The formation of polymers requires the initial step of forming a double hydroxide bridge between monomers. This is illustrated for the dimeric form $Fe_2(OH)_2^{4+}$, the simplest reaction pathway being:

$$\{Fe(H_2O)_5(OH)\}^{2+} + \{Fe(H_2O)_5(OH)\}^{2+} \rightarrow \{Fe_2(H_2O)_8(OH)_2\}^{4+} + 2H_2O \quad \ldots(1.48)$$

The bonding is of the form:

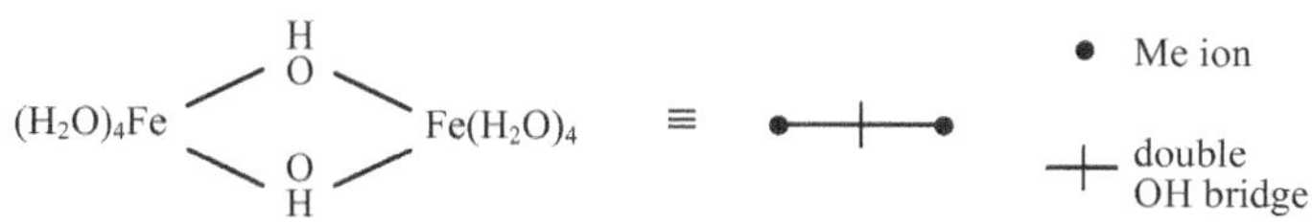

This type of bridging can lead to the evolutionary pathway illustrated in Figure 1.24 in which all or most of the hydroxyls are held in bridged positions. This is known as the *core-links* model (discussed in Section 1.5.3.3).

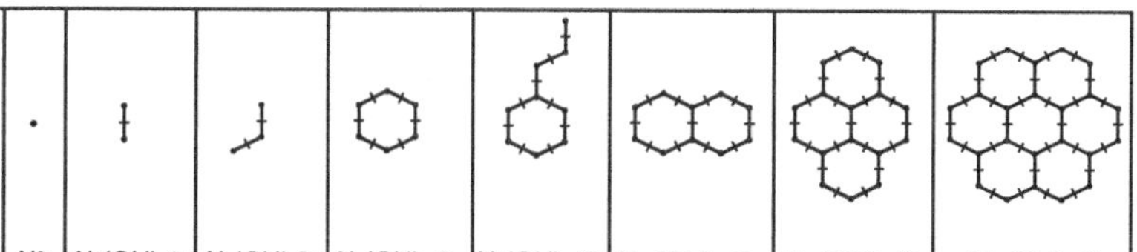

	Al^{3+}	$Al_2(OH)_2^{4+}$	$Al_3(OH)_4^{5+}$	$Al_6(OH)_{12}^{6+}$	$Al_9(OH)_{18}^{9+}$	$Al_{10}(OH)_{22}^{8+}$	$Al_{16}(OH)_{38}^{10+}$	$Al_{24}(OH)_{60}^{12+}$
Average charge / Al	3	2.0	1.67	1.0	1.0	0.80	0.63	0.50
OH / Al ratio	0	1.0	1.33	2.0	2.0	2.20	2.38	2.50

Figure 1.24. Soluble phase species development in core-links model (Stol *et al.*, 1976).

1.5.3.2 Iron polymerisation

An excellent review of the hydrolysis of iron (III) salts is given in Flynn (1984), this providing structural and thermodynamic data together with much detail of polymer formation, kinetics and form. By way of introduction, it is instructive to review features of analysis reported in Gray *et al.* (1995). Figure 1.25 shows a typical neutralisation (titration curve) from van der Woude and de Bruyn (1983) in which a solution of $Fe(NO_3)_3$ was titrated with NaOH.

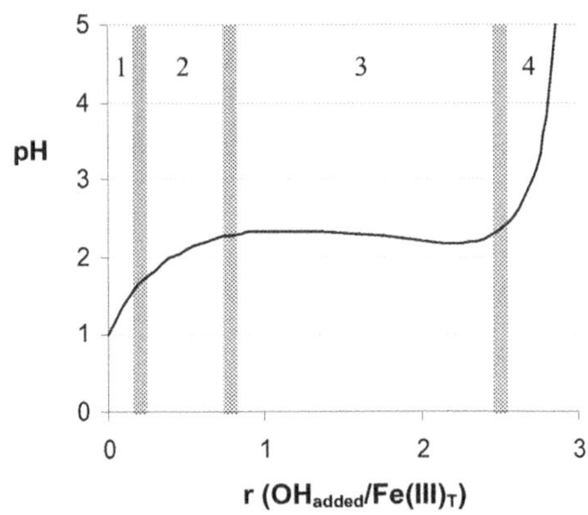

Figure 1.25. Typical base titration curve of $Fe(NO_3)_3$ solution $Fe_T = 6.25 \times 10^{-2}$ M (Gray *et al.*, 1995).

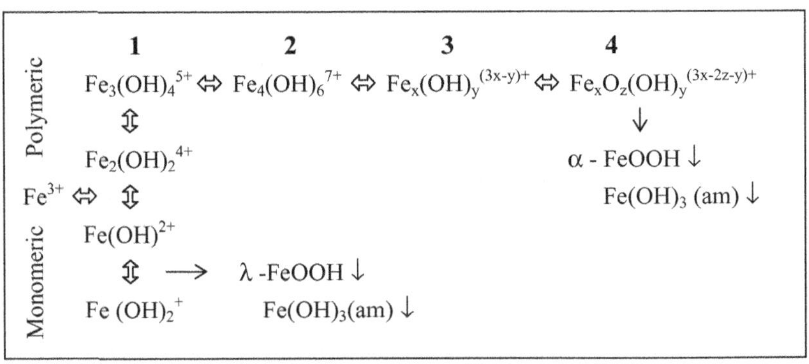

Figure 1.26. Model of iron(III) hydrolysis. The numbers shown at the top of the figure are associated with the zones shown Figure 1.25 (Tang and Stumm, 1987).

The solution pH is plotted as a function of the base addition, denoted by molar ratio r_{OH} = OH_{added}/Fe_T, Fe_T referring to the total iron (III). The trend is subdivided into four zones. In zone 1, with $0 \leq r_{OH} \leq 0.2$, the principal Fe(III) species are monomers and small polymers, such as Fe^{3+}, $FeOH^{2+}$, $Fc_2\ (OII)_2^{4+}$, and $Fe_3(OH)_4^{5+}$; a precipitate is not observed. In zone 2, corresponding to $0.2 \leq r_{OH} \leq 0.8$, polymers larger than the dimer and trimer are formed. The lower boundary of this domain ($r_{OH} = 0.2$) corresponds well with the solubility limit of $Fe(OH)_3$(am). In zone 3 ($0.8 \leq r_{OH} \leq 2.5$) rapid precipitation of $Fe(OH)_3$(am) has been reported and at $r_{OH} = 2.5$ is nearly complete. Continued base addition in zone 4 neutralises the positive charge on the precipitate and produces a very rapid increase in pH. The hydrolysis-polymerisation-precipitation processes associated with base addition to concentrated Fe(III) solutions are shown in Figure 1.26.

Timed calorimetric reactions with ferron reagent are often used to distinguish between the monomers plus dimers and trimers (Fe^a), polymeric iron (Fe^b) and colloidal iron (Fe^c). In the case of slow neutralisation of a 0.3 M $FeCl_3$ solution Gray *et al.* (1995) also employed ultrafiltration to provide insight into species size (see Figure 1.28). Although the concentration is high compared with values used in practice, the plots are instructive in terms of showing the association between species type and approximate size, in response to the degree of neutralisation. One sees evidence of the interchange between the oligomers and the colloidal phase as r_{OH} increases and little evidence of large polymers.

In some respects, these observations tie in with the findings of Bottero *et al.* (1994), who detected dimers and trimers in solution using synchrotron-based X-ray absorption spectroscopy. After 1h, β-FeOOH-like cations (Fe_{24}) were evident, these formed by the coalescence of trimers (Figure 1.27). The Fe_{24} has a size of 1.5 nm and is extremely stable. It appears to resemble a polycation discussed in Tchoubar *et al.* (1991) which acts as a building block in aggregate formation.

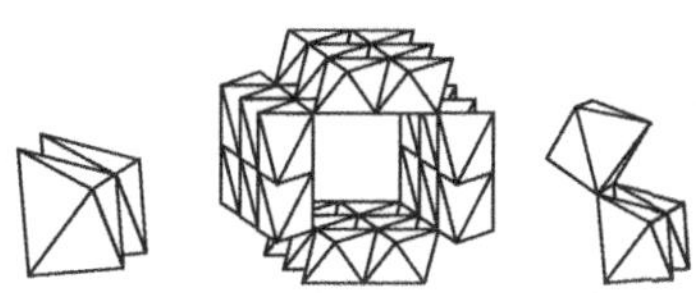

Figure 1.27. Structural models of dimers, trimers and Fe_{24} (adapted from Bottero *et al.*, 1994).

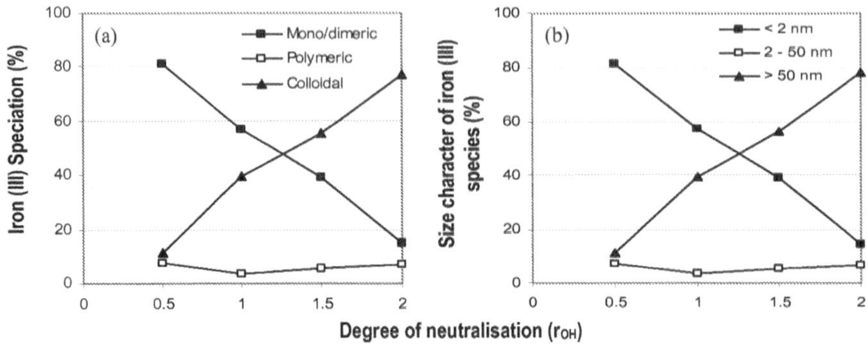

Figure 1.28. Effects of the degree of neutralisation (r_{OH}) on the speciation of polymeric iron chloride preparations (a) based on ferron tests (b) as determined by filtration (Gray *et al.*, 1995).

1.5.3.3 Aluminium polymerisation

In Section 1.5.2 it was evident that the hydrolysis of Al^{3+} and Fe^{3+} were similar with respect to the production of monomers. During the production of polycations, differences emerge. In the case of Al(III) hydrolysis, the combination of octahedral and tetrahedral forms can lead to the formation of Keggin Al_{13} (K-13) whose presence dominates the properties of PACl. The equivalent structural form does not exist in Fe(III) polymerisation. Parallels are strongest when the hydrolysis is spontaneous and rapid (Path I in Figure 1.29), *this being a good approximation of the conditions encountered in practice* when dosing salts such as alum. In these circumstances the formation of K-13 is unlikely. The latter requires highly controlled conditions, termed as *stress hydrolysis* in Tang *et al.* (2004). Between these extremes there exists another pathway in which the evolution of polymeric forms is explained by the *core-links model* (Stol *et al.*, 1976, and Hsu & Bates, 1964). Bi *et al.* (2004) suggests that the different pathways depend on the magnitude of the relative flux of alkali neutralisation defined by:

$$\Phi_r = r_{OH} / t \tag{1.49}$$

in which t is a 'preparation' or aging timescale and $r_{OH} = [OH^-]/[Al^{3+}]$, the mole ratio in the solution. The broad influences of Φ_r are illustrated in Figure 1.29. In path I, the neutralisation rate is so rapid ('high Φ_r') that partial supersaturation occurs and amorphous precipitate is formed. Path II refers to a moderate rate of neutralisation ('moderate Φ_r'), the hydrolysis polymerisation following the core-links model. In path III, the slow neutralisation rate ('small Φ_r') promotes the formation of K-Al_{13}.

Core-links model

The gist of the core-links model or gibbsite fragment scheme is illustrated in Figure 1.24. A key feature of the model is the development of hexamer units which can coalesce to form higher order polymers reflecting the structure of gibbsite. In Figure 1.24 the 'OH/Al ratio' refers to the ligand ratio OH_{bound} /Al within the polycation, this being denoted by the symbol 'n'. At low pH n = 0. As n increases, it is evident that the average charge per Al is reduced, this provides greater opportunities for attachments between sols as the electrostatic interactions are reduced. The net charge refers to the number of ionisable sites on the 'edges' of the complex. For example on the Al_{24} complex there are 18 Al ions on the outer boundary of the cyclic structure of which 6 are fully bonded (3 bridges), the remaining Al ions (with 2 bridges) carrying a single positive charge.

In the domain 2.5 < n < 3.0, further growth in cyclic structures is possible, but there is a tendency for the charges on the edge to be reduced by the attachment of OH ions to the cation. So in this domain, the main Al species are large polymeric Al and sol/gel $Al(OH)_3$. In the case of the polymers, aggregation leading to precipitation occurs when the charge is sufficiently reduced. The transformation from soluble species to the solid includes orientation of the platelets (or sheets) under the influence of hydrogen bonding. At an early stage the aggregates are very disorganised in terms of their stage of growth and orientation and appear as amorphous structures. This is the domain within which there is measurable precipitate. With time and alignment, crystalline structures in the form of gibbsite and bayerite develop.

Generally the core-links model has lacked direct and unequivocal evidence to prove the existence of the transient species, though the existence of the hexamer was verified by Singhal and Keefer (1994) using small angle X-ray scattering.

Keggin Al_{13} (K-13)

The K-Al_{13} polymer refers to the species $AlO_4Al_{12}(OH)_{24}(H_2O)_{12}{}^{7+}$, this representing a building unit for the formation of coagulation structures. It has a central tetrahedral core (AlO_4) surrounded by a 'cage' of 12 octohdredral units gained from monomers, dimers and trimers in solution. There is also a school of thought that its formation depends on the presence of $Al(OH)_4^-$ (Tang *et al.*, 2004). The roughly spherical symmetrical structure of K-Al_{13} is very different from the 'sheet-like' polymeric forms described in the 'core-links' theory. In the range $0.8 < r_{OH} < 2.8$, Al_{13} is regarded as the sole polymeric species.

At r_{OH} = 2, Al_{13} exists as an isolated unit which is stable in solution. For the range 2.3-2.8 the surface progressively loses its charge (within 1 pH unit; see Furrer *et al.*, 1992) leading to the formation of chain structures (length ~ 40 nm) in the first instance and denser clusters with increasing r_{OH} as the contact opportunities increase with charge reduction. At r_{OH} = 2.6 precipitation is evident. Increases in r_{OH} beyond 2.8 see the progressive disappearance of the tetrahedra and the formation of highly polymerised octahedral layers. At r_{OH} =3 the aggregation mechanism is very different, particles being made from a mosaic of platelets and progressing to the short-range order characteristic of bayerite (Bottero *et al.*, 1987).

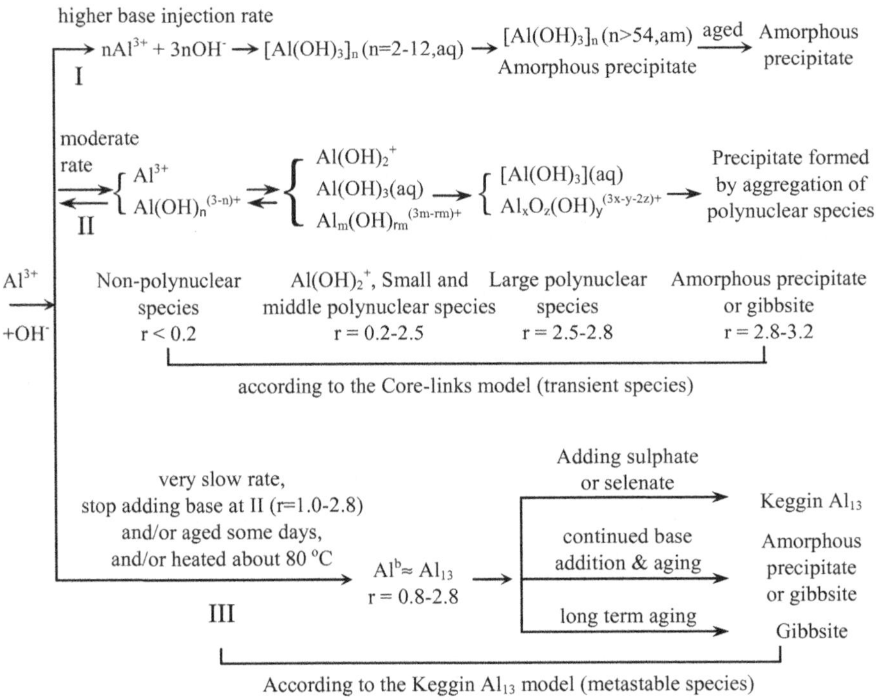

Figure 1.29. The effect of the relative flux of alkali neutralisation (Φ_r) on the course of hydrolysis-polymerisation of Al(III) salts (Bi *et al.*, 2004). Parameters n and m refer to the number of participating molecules and r ($\equiv r_{OH}$) to the solution mole ratio OH/Al; Al^b specifies polymeric species identified in timed calorimetric analysis.

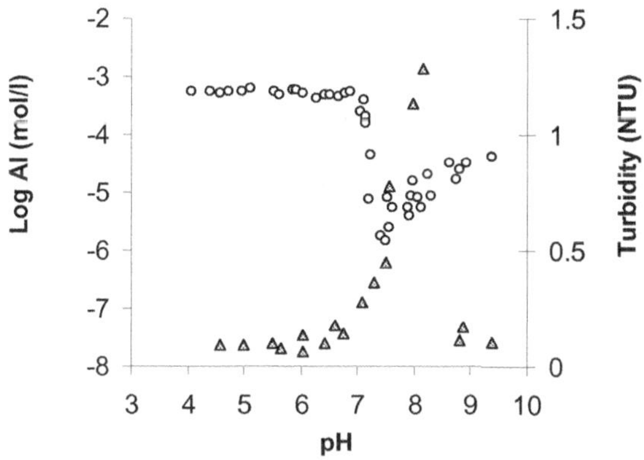

Figure 1.30. Solubility (open circles) and turbidity data (triangles) for PACl added at log Al = -3.3 in deionised water at 25°C. Compiled from Van Benschoten and Edzwald (1990)

Al_{13} appears to be the dominant species in PACl when n is in the range 2 to 2.5 (Tang *et al.*, 2004). In these circumstances one would expect Al_{13} to have a strong influence on the behaviour of the PACl. For example van Benschoten and Edzwald (1990) reported data for PACl shown in Figure 1.30. Although its Al_{13} content was not established, measurements showed that polymerised Al accounted for 90% of the Al content. In the acid domain, it is seen that the solubility of PACl remains constant. This feature together with the lack of increases in turbidity suggests that no hydrolysis has taken place. In the domain pH 6.5 - 8 there is a simultaneous decrease in solubility and increase in turbidity as the structure of PACl is destroyed. Al_{13} responds to pH in a similar way, the critical pH range associated with the destruction of both charge and structure occurring a little lower (see Furrer *et al.*, 1992).

1.5.4 Time factors

Hydrolysis and polymerisation are continuous processes which affect the type and range of species found in solutions and suspensions at different aging times. In the very early stages, Holmes *et al.*, (1968) found that the forward reaction for aluminium hydrolysis associated with Equation (1.46) was on the order of 10^{-5} s (see also Figure 1.21), this being indicative of the transformation timescale of the monomers.

The reactions governing polymerisation are significantly slower. For example the characteristic timescale associated with Equation (1.48) can be represented by:

$$\tau = \frac{1}{[Fe]_T \, k_{11}} \tag{1.50}$$

in which $[Fe]_T$ is the total iron concentration and k_{11} is the forward reaction rate which is estimated to be about 10^3 mol^{-1} dm^3 s^{-1} at 25°C. Examples of this timescale are also shown in Figure 1.21. Generally the production of low molecular weight hydrolysis products (considered as important in destabilisation by the ACN mechanism) is completed within one second (Hahn and Stumm, 1968). In the case of the hydrolysis of Fe(III) solutions over longer periods Flynn (1984) summarises reaction timescales showing their association with the evolution of different structural forms which precede precipitation. Similar data for the hydrolysis of dilute aluminium solutions reported in Smith and Hem (1972). The effects of long-term aging of solutions in which Al_{13} was a dominant polymer at early times have been examined in Parthasarathy and Buffle (1985) and Van Benschoten and Edzwald (1990).

The practical implications of reaction rates and mixing which are significant at the time of dosing and initial dispersion have been examined using a numerical model described in Clark and Srivastava (1993). In this model the dosed aluminium was envisaged as ending up as either amorphous $Al(OH)_3(s)$ or in a polymeric form - the conversion pathways being treated as competing for aluminium atoms. It was shown that poor mixing (longer mixing time scales) favoured conversion into the polymer, whereas good mixing favoured the production of amorphous precipitate. This reflects the gist of the different pathways shown in Figure 1.12.

1.5.5 Charge factors

From the above it is evident that the precipitate is a mixture of neutrally charged materials such as $Al(OH)_3$ and polycations. Beyond the early stages of hydrolysis these tend to be characterised by a zeta potential of the type shown by trend (a) in Figure 1.31; this has an isoelectric point circa pH 9. This type of trend was analysed by Letterman and Asolekar (1990) who suggested that much of the variation could be explained by the pH dependent speciation of charged surface sites on the polynuclear species which were present. The influence of anions - particularly sulphate - can radically alter this picture at longer times as illustrated by trend (b). Trend (c) depicts an intermediate situation, suggesting that the precipitate was dominated by neutrally charged $Al(OH)_3$. Data presented in Amirtharajah and Tambo (1991) show the existence of marked differences in the charge status of aluminium precipitate forms at 2 min and 10 min – the differences being attributed to changes in the polymeric forms with time. These factors are important because they influence the potential of a coagulant to promote destabilisation by charge neutralisation and also affect the nature of the forces which bind precipitates and flocs at different aging times.

The average charge carried by the precipitate is often expressed on the basis of a material in the form $[Mc_p(OH)_q]^{(3p-q)+}$ where q/p is the molar ratio. Sometimes this is represented as an equivalent monomer with p = 1 and q = 2.5, this carrying a charge of 18.5 μeq/mg (Table 1.4).

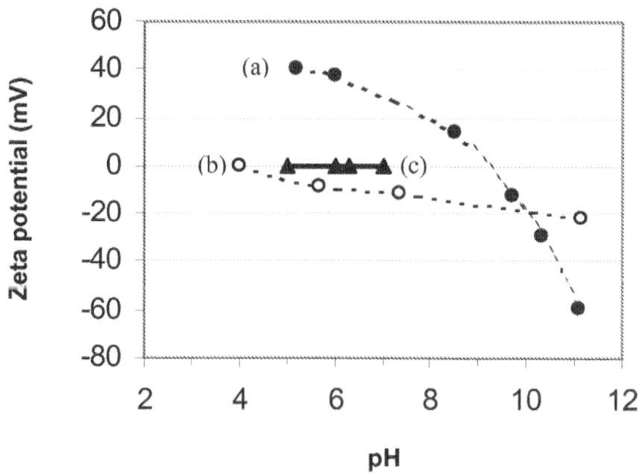

Figure 1.31. Zeta potential of aluminium hydroxide sols under various conditions. Produced from: (a) $Al(NO_3)_2$ at 24 h (Matijevic *et al.*, 1971); (b) $Al_2(SO_4)_3$ at 24 h (Matijevic *et al.*, 1971); (c) $Al_2(SO_4)_3$ immediately after good rapid mixing (Bache *et al.*, 1999).

The corresponding figure for iron is close to 9 μeq/l (definition in Section 1.6). This type of information is very useful in charge neutralisation reactions. For example, Edzwald and Tobiason (1998) note that aquatic humic acid carries a negative charge of about 9 μeq/mg C. If this is offset by a charge of 18.5 μeq/mg Al held on an equivalent monomer (see above), this yields a theoretical stoichiometry for coagulation of about 0.5 mg Al/mg C.

Table 1.4. Soluble aluminium species and positive charge as a function of hydrolysis or basicity (Edzwald and Tobiason, 1998).

Species	OH/Al	% Basicity	Avg. charge per Al	Charge/Al μeq/ mg as Al
Al^{3+}	0	0	3	111
$Al(OH)^{2+}$	1	33	2	74
$Al(OH)_{1.5}{}^{1.5+}$	1.5[(a)]	50	1.5	55.6
$Al(OH)_2{}^{+}$	2	67	1	37
$Al_{13}O_4(OH)_{24}{}^{7+}$	1.25-2.5	42-83	0.54	19.9
$Al(OH)_{2.5}{}^{0.5+}$	2.5[(b)]	83	0.5	18.5

(a) average ligand ratio circa pH 5.5 at 25 °C; (b) corresponding average circa pH 6.5

1.5.6 Surface area

There is much evidence to show that precipitates play an important role in the adsorption of materials from solution. This is often expressed in the form of the BET surface area (Shaw, 1992). Guideline values for a variety of materials are summarised in Table 1.5.

Table 1.5. BET surfaces ($m^2\ g^{-1}$) for various materials (adapted from Masschelein, 1992).

Material	($m^2\ g^{-1}$)	References
Freshly precipitated Fe-hydroxide	300	
α-FeOOH (goethite)	39	Lenoble *et al.* (2002)
Well crystallized β-FeOOH (akaganeite)	36	Bottero *et al.* (1994)
Freshly precipitated Al-hydroxide	40-100, 300*	Letterman & Iyer (1985)*
Well crystallized γ-AlOOH (boehmite)	91	Music *et al.* (1999)
Humic acids from soils	1900	
Kaolinite	10 - 50	
Illite	30 - 80	
Montmorillonite	50 - 150	

From the examples cited, it is evident that freshly precipitated materials generally offer a greater specific surface area (and hence a greater capacity to adsorb) than the well-crystallized forms. This aspect is addressed briefly in Charlet and Manceau (1993, p141) who attribute the greater adsorption capacity of the amorphous forms to their disordered structure. They possess short length octahedral chains /coherent structures (*e.g.* Figure 1.27) and a huge increase in the number of chain terminations compared with the more organised crystalline forms. This results in a greater density of unshared edges available for cation adsorption.

1.5.7 Water binding

One of the salient features of sols and flocs found in water treatment is their large water content, this stemming from the physical chemistry in the early stages of structure formation and aggregation. Licsko (2005) attributes this to hydrogen bonding and suggests that this can arise in two ways. In the first scheme, it is envisaged that water soluble oligomers formed in the early stages of polymerisation grow from forked chains to produce three dimensional structures which enclose relatively large regions of space filled with water (termed as 'closed water'). Within such spaces (on the scale of a few nm), Licsko indicates that the closed water is likely to be characterised by inflexible and irreversible hydrogen bonds. In many respects this is similar to the notion of 'fully immobilised' water noted in Sonntag and Strenge (1987, p142). In the second scheme, Licsko (2005) draws attention to the fact that hydrolysis products are particularly rich in –OH groups with which water molecules can easily form hydrogen bonds. Thus, individual oligomer chains of aluminium or ferric hydroxide are covered by water molecules in the form of flexible and reversible hydrogen bonds.

Whatever pathway is followed, it is inevitable that water will be bound to the metal hydroxides. Liscko goes on to argue that the presence of hydration layers round positively charged sols increases the minimum separation distance of approach of mutually charged sols, thus reducing the impact of electrostatic repulsive forces between the charged surfaces. When the electrostatic forces are sufficiently reduced, relatively weak structures can be formed as a result of the dominance of flexible reversible hydrogen bonds. Because 'bound' water has a more ordered structure than that found in free water, this influences its thermodynamic properties, for example manifesting as a lowering of the freezing point and increasing the viscosity.

1.6 SYNTHETIC ORGANIC POLYMERS

Organic polymers are widely used in water treatment for servicing needs in coagulation, flocculation and dewatering. Their role in destabilisation and flocculation has been briefly reviewed in Section 1.2.8. Here the focus is on some characteristic features rather than their action. A polymer is a chain of small subunits or monomers. The variety, type and number of subunits can be varied to produce materials of different molecular weight. Polymer chains may be linear or branched. For a polymer to be water soluble, it must contain either polar or ionisable groups (*e.g.* –C=O, -OH, $-NH_2$, $-CONH_2$, -COOH, $-SO_3H$). When these exist, the polymer is termed a polyelectrolyte. In general the polymer will also have hydrophobic groups, or an essentially hydrophobic backbone. Depending on the type of ionisable group, a polymer is termed as cationic, anionic, non-ionic (absence of charged groups) or *ampholytic* (containing positive and negative groups). Examples of materials are shown in Figure 1.32.

Charge density (CD) is often expressed as a percentage (mol %) of ionogenic groups relative to all groups of every type (Bolto, 1995). For example, in the case of HPAM shown above:

$$\text{mol\%} = 100f \quad \text{where} \quad f = n/(n+m) \tag{1.51}$$

m and n defining the number of segments of PAA and PAM respectively, the latter carrying n units of charge (*i.e.* n-). Charge density (CD) is also cited as $C\ g^{-1}$ or meq g^{-1} in which 1 eq = 1F = N_Ae = 96487 C mol^{-1} where N_A is the

Avogadro constant ($6.023x10^{23}$) and e is unit charge ($1.6x10^{-19}$ C). When a polymer, such as HPAM, is made up of m segments of uncharged monomer of molecular weight, M_m and n segments of charged polymer of molecular weight M_n each carrying unit charge, the theoretical charge is given by:

$$\frac{meq}{g} = \frac{1000f}{f M_n + (1-f) M_m} \tag{1.52}$$

Examples of the theoretical charge are shown in Table 1.6.

Table 1.6. Theoretical charge density of polymers.

Polymer	Monomeric composition	mol %	meq/g
HPAM	$(C_3H_3O_2)_{0.05}(C_3H_5O)_{0.95}$	5	0.9
PDADMA, Cat-Floc	$C_8H_{16}N$	100	7.9

Charge density will be sensitive to pH in cases in which deprotonation can occur. With regard to PDADMA shown above, deprotonation does not occur, with the consequence that CD is unaffected by pH. Where specific information is missing, Bolto (1995) provides the classification shown in Table 1.7 as a guide to the terminology used in practice.

Table 1.7. Classification of polymeric flocculants by molecular size and charge (Bolto, 1995).

Classification	M (Da)	CD (mol %)
High	$\sim 10^7$	50-100
Medium	$10^5 - 10^6$	~25
Low	$10^4 - 10^5$	~10

The combination of molecular weight and charge density influence the size and conformation of a polymer in a solvent. In isolation, a polymer molecule exists as a form of random coil that might be regarded as a discrete molecular 'particle' whose effective hydrodynamic size (Hogg, 1999) can be estimated from:

$$x_p = 2\xi R_G \tag{1.53}$$

where R_G is the radius of gyration of the molecule (the rms average distance of segments from the centre of mass of the polymer) and ξ is a proportionality coefficient whose value is typically in the range 0.67 to 0.875 (Tanford, 1961).

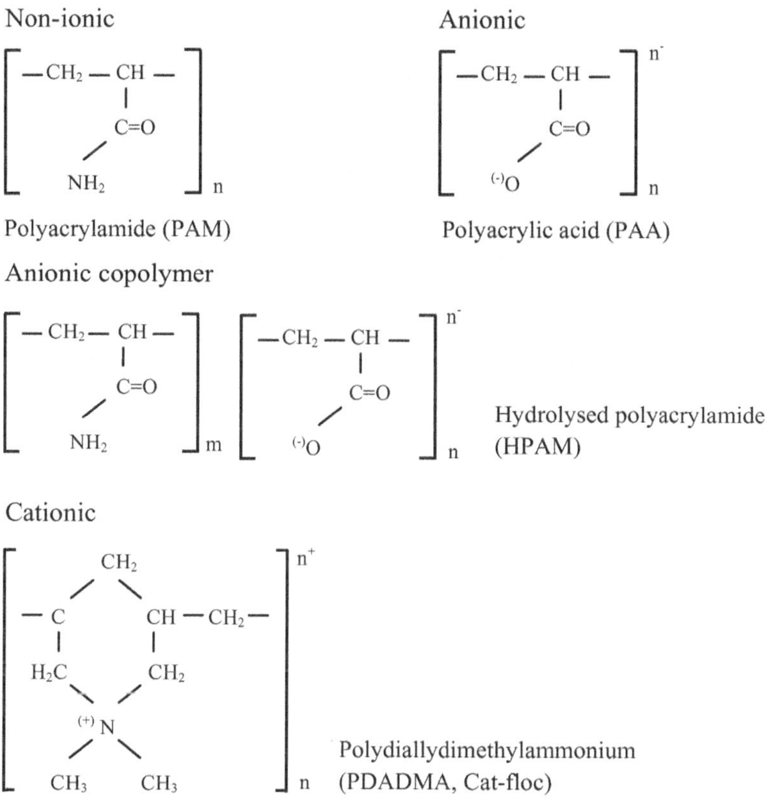

Figure 1.32. Examples of synthetic organic polymers in which the subunits or monomers can be varied to produce materials of different molecular weight and charge (O'Melia, 1972).

The radius of gyration of the molecule can be related to molecular weight through:

$$R_G^2 = \frac{\alpha^2 \beta^2}{6}\left(\frac{M}{M_0}\right) \tag{1.54}$$

where β is the effective bond length in the polymer backbone, M is the polymer molecular weight, M_0 is the segment weight and α is an empirical expansion factor. Data for a variety of polymers in good solvents suggests that:

$$\alpha \approx 0.5 M^{0.09} \tag{1.55}$$

For a polyacrylamide molecule characterised by M $=10^6$ Da, $M_0 = 71$ Da, $\beta = 0.462$ nm, Equations (1.54) and (1.55) indicate $R_G \approx 39$ nm. Substitution into Equation (1.53) shows that the hydrodynamic size $(x_p) \approx 60$ nm.

When a polymer is dissolved in a good solvent (*i.e.* high affinity with polymer) the isolated polymer expands (larger R_G) so as to increase the number of segment-solvent contacts, whereas in a poor solvent the polymer chain contracts. As the charge density increases, the polymer coil becomes more extended as a result of electrostatic repulsion between the segments. Electric interactions are screened by electrolytes. At high ionic strength or low pH an anionic polymer will be tightly coiled, whereas at low ionic strength in neutral or alkaline pH it will be more extended. The reverse occurs with a cationic polymer. If a polymer loses its charge as a result of attachment of small colloids to the active sites, the molecule will tend to contract to its original coil, drawing the attached solids into a coherent floc.

When a polymer is attached to a surface, its conformation will be subject to a wider range of influences, including the charge density of the surface. Yu and Somasundaran (1996) and Besra *et al.* (2004) have each demonstrated that conformation plays a significant role in flocculation and sedimentation as a result of its impact on bridging.

An indication of the effective size of an adsorbed molecule can be gleaned from its adsorption behaviour. The effective size of a molecule can be specified by its 'parking area' σ_p. If one assumes simple square packing at the solid surface (as in Hogg, 1999):

$$\sigma_p \approx x_p^2 \tag{1.56}$$

the maximum adsorption density (Γ_m) would be:

$$\Gamma_m \approx \frac{M}{N_A x_p^2} \tag{1.57}$$

Referring to the example shown above for PAM, Equation (1.57) predicts a maximum adsorption density of about 2.3 mg m^{-2}. Conversely, if one has measured Γ_m from adsorption isotherms, the molecular size can be estimated using Equation (1.57) *e.g.* Besra *et al.* (2004). Using the same type of approach, Parazak *et al.* (1988) demonstrated the significance of hydrophobic interactions for promoting aggregation.

1.7 SOLIDS PRODUCTION AND COMPOSITION

For planning purposes, a number of formulae have been proposed for estimating the dry solids production, whose magnitude and composition has a major bearing on the design and operation of solid-liquid separation processes. The formula shown below is cited in Gregory and Dillon (2002):

Sludge solids* (mg/l) = 2 x turbidity removed from water (NTU)
+ 0.2 x colour removed (Hazen)
+ 2.9 x aluminium precipitated (mg Al/l)
+ 1.9 x iron precipitated (mg Fe/l)
+ 1.6 x manganese precipitated (mg Mn/l)
+ non-soluble lime solids etc., precipitated (mg/l)
...(1.58)

* *refers to suspended solids production potential for water being treated*

Inspection of Equation (1.58) shows that the solids are drawn from three sources, these being (i) raw water, (ii) coagulant, and (iii) other added chemicals. Although rudimentary, Equation (1.58) provides insight into the relative contribution from these sources. On cost grounds, a general aim is to minimise the materials added. It is also found that it is the materials added, rather those gained from (i) which can dominate the behaviour of the mix. In the case of (ii), this should come as no surprise as the coagulant products because of their relatively large surface area (see Table 1.5) and surface properties.

Before discussing this further, some caution needs to be exercised with respect to Equation (1.58). First, when dealing with an operational works, the solids production is easily measured *e.g.* Bache *et al.* (2003). Second, it grossly simplifies the chemistry. For example, the conversions from the metal ions is based on the molar ratio $[Me(OH)_q]/[Me^{q+}]$. What takes place in practice is far more complex. For example, Figure 1.33(a) illustrates the conversion from an $Al_2(SO_4)_3$ solution in the absence of other materials apart from those used to adjust pH. For the Al concentrations cited, the solids liberated collapse onto a single trend irrespective of pH. This has a maximum at about pH 4.8 with a conversion ratio of 3.7. Some of the 'extra' precipitate can be attributed to conversions from the sulphate.

When water contains colour, other complexities arise. For example, Bache *et al.*, (2003) found that the conversion shown in Equation (1.58) overestimates the solids release, a better representation being 0.1 x true colour. Hossain and Bache (1991) measured the solids release from a low turbidity, soft coloured water coagulated with alum. After allowing for the solids release from colour and turbidity, the estimated conversion from the Al ion is shown in

Figure 1.33(b). This has a very different shape from the trend shown in Figure 1.33(a), the striking feature being a narrowing of the pH range associated with the solids release. Maximum conversion again occurs at around pH 4.8 with a conversion ratio exceeding 4. At lower pH, data collapses to a single trend, whereas at higher pH the lack of collapse points to a dose-related interaction between the colour, turbidity and coagulant. In the context of Figure 1.33(b), Bache and Hossain (1991) found that the optimum pH (for both coagulation and floc properties) corresponded to that associated with the maximum solids release (Section 3.2.4).

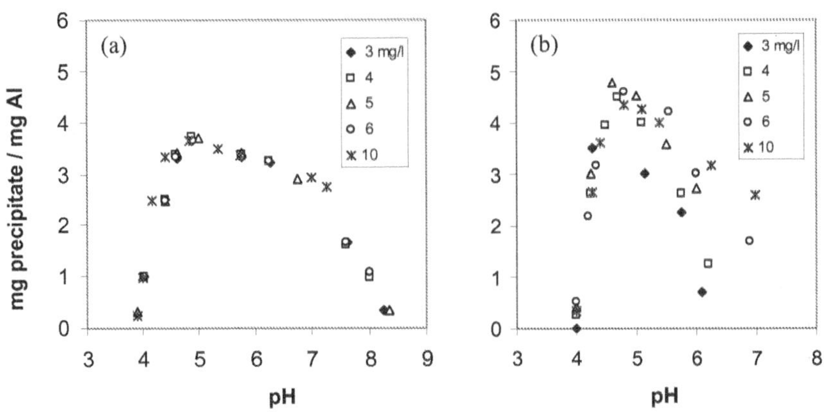

Figure 1.33. Scaled solids release from (a) a distilled water base (b) raw-water base (turbidity ~ 2 NTU, colour ~ 100 Hazen) based on data in Hossain and Bache (1991).

One of the major determinants of floc character is the solids volume concentration, this depending on floc size and the solids density (Section 2.1.3). The effective value of the solids density depends on the materials present. A material balance shows that the component densities are linked by the formula:

$$\frac{1}{\rho_s} = \frac{1}{\rho_p} + f\left(\frac{1}{\rho_c} - \frac{1}{\rho_p}\right) \tag{1.59}$$

in which ρ_s, ρ_p, ρ_c signify the dry densities of the floc solids, the raw-water primary particles and coagulant products respectively, f referring to the fraction of coagulant products within the dry mass.

Hossain and Bache (1991) used this model to characterise the solids arising from the coagulation of a low-turbidity coloured water. Composition (*i.e.* fraction f) was varied by adjusting the coagulant dose. The study showed $\rho_p = 1750 \pm 40$ kg m^{-3} (mainly humic) and $\rho_c = 2530 \pm 50$ kg m^{-3} (corresponding to $Al(OH)_3$ (s)). It was found that f ≈ 0.5 at the optimum coagulation conditions. This figure is consistent with Warden (1984) who indicates that f ~ 0.45 for alum sludges derived from reservoir waters (considered to be of similar type) and ~ 0.11 for sludges gained from lowland waters.

1.8 REFERENCES

Adler, P.M. (1981) Heterocoagulation in shear flow. *J. Colloid Interface Sci.*, 83(1), 106-115.

Amirtharajah, A. and Tambo, N. (1991) Mixing in water treatment. In: *Mixing in Coagulation and Flocculation* (Eds Amirtharajah,A., Clark, M.M. and Trussell, R.R.) American Water Works Research Foundation, Denver, Co.

Bache, D. H. and Hossain, M. D. (1991) Optimum coagulation conditions for coloured water in terms of floc properties. *J. Water SRT-Aqua*, 40, 170-178

Bache, D.H., Johnson, C., Papavasilopoulos, E., Rasool, E and McGilligan, J.F. (1999) Sweep coagulation: structures, mechanism and practice. *J. Water SRT-Aqua*, 48(5), 201-210.

Bache, D.H. and Rasool, E (2001) Characteristics of turbulence in a multigrid mixer. *Chem. Eng. J.*, 83, 67-78.

Bache, D.H., Papavasilopoulos, E.N., Rasool, E., Zhao, Y.Q. and McGilligan, J.F. (2003). Polymers in alum sludge dewatering: developments and control. *J. CIWEM*, 17 (2), 106-110.

Baldyga, J and Bourne, J. R. (1984) A fluid mechnical approach to turbulent mixing and chemical reaction. Part II. Micromixing in the light of tubulence theory. *Chem. Eng. Commun.*, 28, 243-258.

Baldyga, J and Bourne, J. R. (1995) Interpretation of turbulent mixing using fractals and multifractals. *Chem Eng. Sci.*, 50(3), 381-400

Behrens, S. H. and Borkovec, M. (2000) Influence of the secondary interaction energy minimum on the early stages of colloidal aggregation. *J. Colloid Interface Sci.*, 225, 460-465.

Besra, L., Sengupta, D.K., Roy, S.K. and Ay, P. (2004) Influence of polymer adsorption and conformation on flocculation and dewatering of kaolin suspension. *Sep. Purif. Tech.*, 37, 231-246.

Bi, S., Wang, C., Cao, Q. and Zhang, C. (2004) Studies on the mechnism of hydrolysis and polymerisation of aluminium salts in aqueous solution: correlations between the "Core-links" model and the "Cage-like" Keggin-Al_{13} model. *Coordination Chem. Rev.*, 248, 441-455.

Bolto, B.A. (1995) Soluble polymers in water purification. *Prog. Polym. Sci.*, 20, 987-1041.

Bottero, J.Y., Axelos, M., Tchoubar, D., Cases, J.M., Fripiat, J.J. and Fiessinger, F. (1987) Mechanism of formation of aluminium trihydroxide from Keggin Al_{13} polymers. *J. Colloid Interface Sci.*, 117(1), 47-57.

Bottero, J.Y., Manceau, A., Villieras, F. and Tchoubar, D. (1994) Structure and mechanisms of formation of FeOOH(Cl) polymers, *Langmuir*, 10, 316-319.

Brodkey, R.S. (1967) *Phenomena of Fluid Motions*. Addison-Wesley, Reading, Mass.

Charlet, L. and Manceau, A. (1993) Structure, formation, and reactivity of hydrous oxide particles: insights from x-ray adsorption spectroscopy. In: *Environmental Particles*. Vol. 2 (Eds Buffle, J. and van Leeuwen, H.P.) Lewis Publishers, Florida.

Clark, M.M. and Srivasta, R.M. (1993) Mixing and aluminium precipitation. *Environ. Sci. Technol.*, 27, 2181-2189.

Corrsin, S. (1964) The isotropic turbulent mixer: Part II. Arbitrary Schmidt Number. *AIChE J.*, 10(6), 870-877.

David, R. and Clark, M.M. (1991) Micromixing models and application to aluminium netralization precipitation reactions. In: *Mixing in Coagulation and Flocculation* (Eds Amirtharajah,A., Clark, M.M. and Trussel, R.R.) American Water Works Research Foundation, Denver, Co.

Dentel, S. K. (1988) Application of the precipitation-charge neutralization model of coagulation. *Environ. Sci. Technol.*, 22(7), 825-832.

Dentel, S. K (1991) Coagulant control in water treatment. *Critical Reviews in Environmental Control*, 21(1), 41-135.

Derjaguin, B.V. and Landau, L.D. (1941) Theory of the stablility of strongly charged lyophobic sols and the adhesion of strongly charged particles in solutions of electrolytes. *Acta Physicochim. URSS*, 14, 733-762.

Duan, J. and Gregory, J. (2003) Coagulation of hydrolysing metal salts. *Adv. Colloid Interface Sci.*, 100-102, 475-502.

Dymaczewski, Z., Kempa, E.S. and Sozanski, M.M. (1997) Coagulation as a stucture-forming separation process in water and wastewater treatment. *Water. Sci. Tech.*. 36(4), 25-32.

Edwards, M.F. (1985) Mixing of low-viscosity liquids in stirred tanks. In: *Mixing in the Process Industries* (Eds Harnby, N., Edwards, M.F. and Nienow, A.W.) Butterworths, London.

Edzwald, J.K. and Tobiason, J.E. (1998) Enhanced versus optimised multiple objective coagulation. In: *Chemical Water and Wastewater Treatment V.* (Eds Hahn, H.H., Hoffmann, E. and Odegaard, H.) Springer, Berlin. p113-124.

Elimelech, M, Gregory, J., Jia, X. and Williams, R.A. (1995) *Particle Deposition and Aggregation*. Butterworth-Heinemann, Oxford.

Flesch, J.C., Spicer, P.T. and Pratsinis, S.E. (1999) Laminar and turbulent shear-induced flocculation of fractal aggregates. *AIChE J.*, 45(5), 1114-1124.

Flynn, C.M. (1984) Hydrolysis of inorganic salts. *Chem. Rev.*, 84, 31-41.

Fuchs, N. (1934) Über die Stabilität und Aufladung der Aerosole. *Z. Physik*, 89, 736-743.

Furrer, F., Ludwig, C. and Schindler, P.W. (1992) On the chemistry of Keggin Al_{13} polymer. *J. Colloid Interface Sci.*, 149(1), 56-67.

Godfrey, J. C. and Amirtharajah, A. (1991) Mixing in liquids. In: *Mixing in Coagulation and Flocculation* (Eds Amirtharajah,A., Clark, M.M. and Trussell, R.R.) American Water Works Research Foundation, Denver, Co.

Gray, K.A., Yao, C. and O'Melia, C.R. (1995) Inorganic metal polymers: preparation and characterisation. *J. AWWA*, 87(4), 136-148.

Gregory, J. (1989) Fundamentals of flocculation. *CRC Critical Reviews in Environmental Control*, 19(3), 185-230.

Gregory, R. and Dillon, G. (2002) Minimising sludge production at water-treatment plants. *J. CIWEM*, 16(3), 174-179.

Hahn, H.H. and Stumm, W. (1968) Kinetics of coagulation with hydrolysed Al(III): the rate-determining step. *J. Colloid. Interface Sci.*, 28(1), 134-144.

Hall, T (1997) *Water Treatment Processes and Practices.* 2nd ed.,WRc, Swindon, UK.

Hamaker, H.C. (1937) The London-van der Waals attraction between spherical particles. *Physica* 4, 1058-1072.

Han, M. and Lawler, D. F. (1992) The (relative) insignificance of G in flocculation. *J. AWWA*, 84(10), 79-91.

Healy, T.W. and La Mer, V.K. (1964) Energetics of flocculation and redispersion by polymers. *J. Colloid Sci.*, 19, 323-332.

Herweijer, J.A. (1995) The Small-Scale Structure of Turbulence. Technische Universiteit Eindhoven, Eindhoven. ISBN 90-386-0137-9.

Hinze, J.O. (1987) *Turbulence.* 2nd Edition, McGraw-Hill, New York.

Holmes, L. P., Cole, D. L., Eyring, E. M. (1968) Kinetics of aluminium hydrolysis in dilute solutions. *J. Phys. Chem.*, 72(1), 301-304.

Hogg, R. (1999) The role of polymer adsorption kinetics in flocculation. *Colloids Surfaces A: Physicochem. Eng. Aspects*, 146, 253-263.

Hossain, M. D. and Bache, D.H. (1991) Composition of alum flocs derived from a coloured low-turbidity water. *J. Water SRT-Aqua*, 40, 298-303.

Hsu, P.H. and Bates, T.F. (1964) Formation of X-ray amorphous and chrystalline and aluminium hydroxides. *Mineralogical Magazine*, 33, 749-768.

Huber, E. and Frost, M. (1998) Light scattering by small particles. *J. Water SRT-Aqua*, 47(2), 87-94.

Hunter, R.J. (2001) *Foundations of Colloid Science.* Second Edition. Oxford University Press, Oxford, UK.

Kramer, T.A. and Clark, M.M. (1999) Incorporation of aggregate breakup in the simulation of orthokinetic coagulation. *J. Colloid Interface Sci.,* 216, 116-126.

Lenoble, V., Bouros, O., Deluchat, V., Serpaud, B. and Bollinger, J.C. (2002) Arsenic adsorption onto pillared clays and iron oxides. *J. Colloid Interface Sci.*, 255, 52-58.

Letterman, R.D. and Asolekar, S.R. (1990) Surface ionisation of polynuclear species in Al(III) hydrolysis-II. Conditional equilibrium model. *Water. Res.*, 24(8), 941-948.

Letterman, R. D. and Iyer, D. R. (1985) Modelling the effects of hydrolyzed aluminium and solution chemistry on flocculation kinetics. *Environ. Sci. Technol.*, 19(8), 673-681.

Levich, V.G. (1962) *Physicochemical Hydrodynamics.* Prentice Hall. Englewood Cliffs, US.

Licsko, I. (2005) Coagulation considerations – the main processes. Proceedings of IWA International Conference on Particle Separation 2005, June 1-3, Seoul, Korea, 223-230.

Masschelein, W.J. (1992) *Unit Processes Drinking Water Treatment,* Marcel Dekker, New York.

Matijevic, E., Mangravite, F.J. and Cassell, E.A. (1971) Stability of colloidal silica IV. The silica-alumina system. *J. Colloid Interface Sci.*, 33(4), 560-568.

Music, S. Dragcevic, D. and Popovic, S. (1999) Hydrothermal crystallization of boehmite from freshly precipitated aluminium hydroxide. *Materials Letters* 40, 269-274.

O'Melia, C. R, (1972) Coagulation and Flocculation. In: *Physicochemical Processes for Water Quality Control.* (Ed., Weber, W.J.) Wiley-Interscience, New York

Parazak, D.P., Burkhardt, C.W., McCarthy, K.J. and Stehlin, M.P. (1988) Hydrophobic flocculation. *J Colloid Interface Sci.*, 123(1), 59-72.

Parthasarathy, N and Buffle, J. (1985) Study of polymeric aluminium(III) hydroxide solutions for application in waste water treatment. *Water Res.*, 19(1), 25-36.

Reerink, H. and Overbeek, J. Th. G. (1954) The rate of coagulation as a measure of the stability of silver iodide sols. *Discuss. Faraday Soc.*, 42, 267-275.

Sawyer, C.N., McCarty, P.L. and Parkin, G.F. (1994) *Chemistry for Environmental Engineering.* Forth edition. McGraw Hill, New York.

Shaw. D.J. (1992) *Introduction to Colloid and Surface Chemistry*. Fourth Edition. Butterworth-Heinemann, Oxford.

Singhal, A.O. and Keefer, K.D. (1994) A study of aluminium speciation in aluminium chloride solutions by small angle x-ray and ^{27}Al NMR. *J. Mater. Res.*, 9(8), 1973-1983.

Smith, R.W. and Hem, J.D. (1972) *Effect of aging on aluminium hydroxide complexes in dilute aqueous solutions*. U.S. Geol. Surv. Water Supply Paper 1827-D. U. S. Government Printing Office, Washington.

Sonntag, H. (1993) Coagulation kinetics. In: *Coagulation and Flocculation, Theory and Applications* (Ed., Dobiáš, B.) Marcel Dekker, New York.

Sonntag, H. and Strenge, K. (1987) *Coagulation kinetics and structure formation*. Plenum, New York.

Stol, R.J., Van Helden, A.K. and De Bruyn, P.L. (1976) Hydrolysis-precipitation studies of aluminium (III) solutions. II. A kinetic study and model. *J. Colloid Interface Sci.*, 57, 115-131.

Stumm, W. (1992) *Chemistry of the Solid-Water Interface*. John Wiley, New York.

Stumm, W. and O'Melia, C. R., (1968) Stoichiometry of coagulation. *J. Amer. Water Works Assoc.*, 60, 514-539.

Tambo, N and Hozumi, H. (1979) Physical aspect of flocculation process – II. Contact flocculation. *Water Res.*, 13, 441-448.

Tang, H. X. and Stumm, W. (1987) The coagulation behaviors of Fe(III) polymeric species – 1. Preformed polymers by base addition. *Water Res.*, 21(1), 115-121.

Tang, H. X., Wang, D.S. and Xu, Y. (2004) Optimization of the concepts for polyaluminium species. In: *Chemical Water and Wastewater Treatment VIII* (Eds. Hahn, H.H., Hoffmann, E. and Odegaard, H.), IWA Publishing, London, p139-149.

Tanford, C. (1961) *Physical Chemistry of Macromolecules*. Wiley, New York.

Tennekes, H. and Lumley, J.L. (1994) *A first course in turbulence*. MIT Press, Cambridge, MA.

Tchoubar, D., Bottero, J.Y., Quienne, P. and Arnaud, M. (1991) Partial hydrolysis of ferric chloride salt. Structural investigation by photon-correlation spectroscopy and small-angle X-ray scattering. *Langmuir*, 7, 398-402.

van der Woude, J.H.A. and de Bruyn, P.L. (1983) Formation of colloidal dispersions from supersaturated iron (III) nitrate solutions. I. Precipitation of amorphous iron hydroxide. *Colloids and Surfaces*, 8, 55-78.

Van Benschoten, J.E. and Edzwald, J.K. (1990) Chemical aspects of coagulation using aluminium salts – 1. Hydrolytic reactions of alum and polyaluminium chloride. *Water Res.*, 24, 1519-1256.

Verwey, E.J.W. and Overbeek, J.Th.G. (1948) *Theory of the Stability of Lyophobic Colloids*. Elsevier, Amsterdam.

Vivaldo-Lima, E., Wood, P.E., Hamielec, A.E. and Penlidis, A. (1998) Calculation of the particle size distribution in suspension polymerisation using a compartment-mixing model. *Can. J. Chem. Eng.*, 76, 496-505.

Von Smoluchowski, M. (1917) Versuch einer mathematischen Theorie der Koagulationskinetic kolloider Lösungen. *Z. Phys. Chem.*, 92, 129-168.

Warden, J.H. (1984) Thickening and dewatering of hydroxide sludges. In: Gregory, J. (Ed.) *Solid Liquid Separation*. Ellis Horwood, Chichester, UK.

Yu, X and Somasundaran, P. (1996) Kinetics of polymer conformational changes and its role in flocculation. *J. Colloid Interface Sci.*, 178, 770-774.

Section 2 Characterisation

2.1 PARTICLE STRUCTURES

2.1.1 Introduction

Many studies from the 1960's and 1970's *e.g.* Lagvanker & Gemmell (1968) and Tambo & Watanabe (1979), have demonstrated that properties such as the floc density-size relationship follows a power law dependence of the form l^{γ} in which l is a floc length scale and γ (> 0) is a constant associated with a particular size range and conditions. At much the same time, the science of fractal geometry (*e.g.* Mandelbrot, 1977) emerged as a means of describing the form of objects (or the patterns of data in space) which do not comply with the scaling relationships predicted by Euclidean geometry for similar shaped entities of differing size. The vital feature of fractal behaviour is the preservation of some characteristic pattern at different length scales. With advances in measurement techniques, theory and numerical simulation studies, there has been an upsurge in studies making use of this approach for describing spatial patterns in data.

In this section, the intention is to outline the framework that is often used to describe fractal objects and to illustrate how it is applied to clusters. This description underpins the representation of floc structure and its impact on physical properties. Where there exists a single power law dependence between a characteristic property and size *i.e. monofractal* behaviour, this might be attributed to the existence of a fractal structure in the component solids. Alternatively, the property may display trends, which can be fitted by more than one value of γ across the size range (see Figure 2.3) - perhaps indicating a *multifractal* structure. However, it must be stressed that the existence of a property-size dependence following a power law does not necessarily imply the existence of a fractal structure, because the power law dependence may arise from physical factors quite unconnected with the structure. To discriminate, one must understand the controlling factors.

2.1.2 Fractal dimensions

When one has a set of data points embedded in space, the fractal dimension is a measure which conveys information about the way a pattern of behaviour is

ISBN: 9781789065442. Published by IWA Publishing, London, UK.

replicated at different size scales - particularly at the limit of vanishingly small size. Depending on the statistical properties under scrutiny, there are many types of fractal dimension that can be used to describe the set. Those in common use include the box-counting dimension (D_B), the information dimension (D_I) and the correlation dimension (D_C), which represent particular cases of multifractal behaviour (*e.g.* Moon, 1992).

2.1.2.1 Box-counting dimension

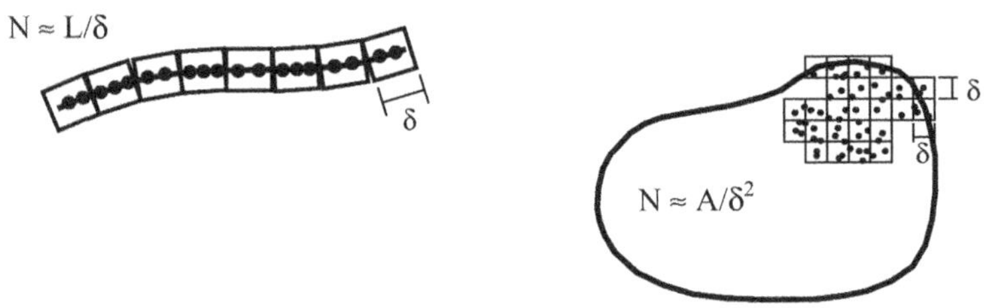

Figure 2.1. Illustrating how spatial distributions can be covered by a network of 'hypercubes' for counting purposes (based on Moon, 1992).

If an object with *hypervolume* V^* (a general term used to represent a length, area, volume or a hypervolume with a non-integer or *fractal* space dimension) is covered by a network of *hypercubes* of side length δ such as illustrated in Figure 2.1, the number of hypercubes, N(δ), required to cover or measure the size or capacity of V^*, can be expressed by the general relationship:

$$N(\delta)\delta^D = V^* \tag{2.1}$$

In Equation (2.1), D (≡ D_B) refers to the *capacity or box-counting* dimension. For example, when the hypervolume refers to a distribution of points along a line of length L, then N is identified via the relationship $N\delta^1 \approx L$. In this example the identity D = 1 is necessitated by the underlying dimension of the reference line. The measurement is an approximation because the number of units of δ^1, does not necessarily yield an exact evaluation of L. Indeed, one may vary the length of δ in order to evaluate L. Equation (2.1) can also be written in the form:

$$\log N = D_B \log(1/\delta) + \log V^* \tag{2.2}$$

From a plot of log N(δ) against log (1/δ), the slope is represented by:

$$D_B = \lim_{\delta \to 0} \frac{\log N(\delta)}{\log(1/\delta)} \tag{2.3}$$

This represents a formal definition of D_B and is one means by which D can be evaluated in terms of a probing length δ. Mathematically, there are more rigorous ways of specifying D in terms of the limiting behaviour of a test function (*e.g.* Feder (1988), p 12-15).

2.1.2.2 Information dimension

A limitation of the box-counting dimension D_B is that it provides no insight into the distribution of points within the covering cells. This is addressed by the information entropy (Shannon) dimension, D_I, which reflects the probability of finding a point in the cell.

To estimate the information dimension, one counts the number of points N_i in each of the N cells and determines the probability P_i of finding a point in that cell where:

$$P_i = \frac{N_i}{N_0}, \quad \sum_{i=1}^{N} P_i = 1 \tag{2.4}$$

in which N_0 is the total number of points in the set. Following the description provided in Moon (1992), the *information entropy* is defined by the expression:

$$I(\delta) = -\sum^{N} P_i \log P_i \tag{2.5}$$

When δ is small, it can be shown that I(δ) behaves as:

$$I \approx D_I \log(1/\delta) \tag{2.6}$$

allowing D_I to be defined by:

$$D_I = \lim_{\delta \to 0} \frac{I(\delta)}{\log(1/\delta)} = \lim_{\delta \to 0} \frac{\sum P_i \log P_i}{\log \delta} \tag{2.7}$$

The information entropy is a measure of the unpredictability in a system. For example, if all the points were located in one cell, say with $P_1 = 1$ and $P_i = 0$ for $i \neq 1$, then $I = 1 \log(1) = 0$ and $D_I = 0$. Here, the value $I = 0$ corresponds to a state of maximum predictability. At the other extreme, if all the points are equally distributed across the cells, $P_i = 1/N$, $I = \log N$, showing that $D_I = D_B$; in this case the information entropy is at its maximum value.

2.1.2.3 *Correlation dimension*

The correlation dimension centres on information gained from a probing hypersphere of radius r. It accounts for the probability of finding two points in the measurement cell. If one has a set of N points $\{\mathbf{x}_i\}$ in the phase space, the distance between a pair of points can be expressed by $s_{ij} = |\mathbf{x}_i - \mathbf{x}_j|$. The probing hyposphere is centred on each point in the set and counts the number of points within the hyposphere. A correlation function C(r) is defined by the summation (based on Balankin (1996) and Moon, 1992):

$$C(r) = \frac{1}{N(N-1)} \sum \begin{pmatrix} \text{number of pairs (i, j)} \\ \text{with distance } s_{ij} < r \end{pmatrix} \tag{2.8}$$

Where there is fractal behaviour, the correlation function exhibits a power law dependence on r as $r \to 0$ in accord with:

$$\lim_{r \to 0} C(r) = ar^{D_C} \tag{2.9}$$

From the slope of log C(r) versus log r plot, one may identify the correlation dimension D_C which has the formal definition:

$$D_C = \lim_{r \to 0} \frac{\log C(r)}{\log r} \tag{2.10}$$

2.1.2.4 *Relationship between fractal dimensions*

The dimensions described above form part of a generalised collection of dimensions, known as Rényi dimensions (see below). Moon (1992) reports that D_C forms the lower bound of D_I, which in turn forms a lower bound to D_B *i.e.*

$$D_C \le D_I \le D_B \tag{2.11}$$

When points are uniformly distributed across the probing cells, $D_C = D_I = D_B$, showing a single fractal dimension. Conversely, if a set is monofractal, then it implies that the above dimensions are equal.

2.1.2.5 *Multifractals*

Multifractal structures arise when the data points are not uniformly distributed across the counting cells, the set being characterised by a suite of fractal dimensions (D_q) reflecting differing scaling properties of subsets. These are identified by weighting the cell probabilities using the form P_i^q in which q refers to a moment of a probability, q taking values from $-\infty$ to $+\infty$. When $q \rightarrow -\infty$ a bias is given to the cells with the lowest probability, whereas $q \rightarrow \infty$ gives bias to the cells with the highest P_i, from this the associated dimensions D_q (*Rényi dimensions e.g.* Balankin, 1996) are defined by the following pair in association with Equation (2.4):

$$D_q = \lim_{\delta \to 0} \frac{I_q(\delta)}{\log \delta} \tag{2.12}$$

$$I_q = \frac{1}{q-1} \log \left(\sum_{i=1}^{N(\delta)} P_i^q(\delta) \right) \tag{2.13}$$

Here I_q is the generalised entropy of order q. When $q = 0$, $\Sigma P_i^q = N(\delta)$ indicating that $D_0 \equiv D_B$, the box-counting dimension. At $q = 1$, $D_1 \equiv D_I$ (the information entropy as defined by Equation (2.5). The generalised dimension D_2 corresponds to the correlation dimension, D_C.

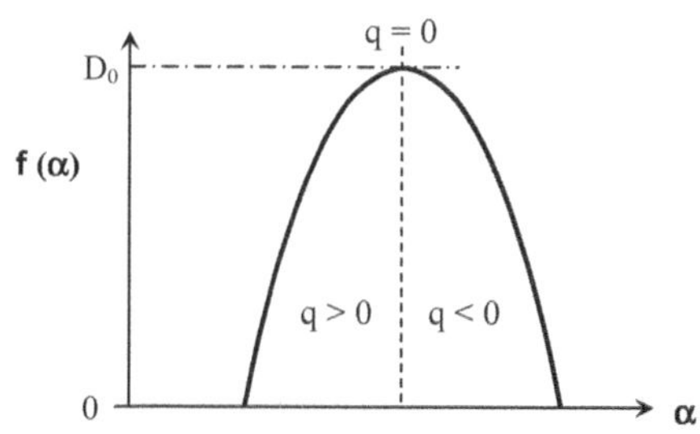

Figure 2.2. Singularity spectrum for a multifractal structure (based on Feder, 1988).

If one treats q as a continuous variable, data from Equations (2.12) and (2.13) can be translated into a *singularity spectrum* f(α) in which α is the Lipschitz-Holder exponent or *crowding index* of the data. The logic behind this representation is described in Feder (1988), or may be calculated directly from experimental data using the procedures described in Chabra and Jensen (1989). The magnitude of f(α) depends on q and may be interpreted as the fractal dimension of a subset that dominates the value of ΣP_i^q in Equation (2.13).

Overall f(α) lies in the interval $0 \leq f(\alpha) \leq D_0$. Plots of the singularity spectrum based on knowledge of D_0 generally possess a roughly parabolic shape (*e.g.* Figure 2.2) with D_0 as the maximum. It serves as a useful means of representing multifractal properties, and provides a deeper insight into similarity behaviour across a size range. Grout *et al.* (1998) make use of the Chabra and Jensen approach to represent particles size distributions found in soil; from this they demonstrated that data gained from different soil horizons exhibited the same f(α) plot *i.e.* displaying a common multifractal characteristic.

At present, there has been very little exploitation of this approach. It should be regarded as a useful framework for describing the different facets of structure. For example, it was used in Balankin (1996) as a means of describing the elastic behaviour of materials with a multifractal structure (Section 2.3.6.1).

2.1.3 Clusters

2.1.3.1 Descriptors of structure

Here we focus on clusters representing an idealised floc in which the encased volume is set in three-dimensional space. By analogy, if we were to examine the cross-section of a floc, its encased area exists in two-dimensional space. In contrast, the statistical distribution of points or particles making up the cluster will have their own characteristic dimension (D) which is likely to be different from the prevailing Euclidean dimension.

For a cluster of radius R containing N particles of radius a, the number-size dependence can be expressed by:

$$N = A'\left(\frac{R}{a}\right)^{D} \tag{2.14}$$

in which the corresponding volume averaged cluster volume density (volume solids/cluster volume) is:

$$\phi_R = A'\left(\frac{R}{a}\right)^{D-3} \tag{2.15}$$

where A′ is a packing factor. Following the methodology described in Sonntag and Russel (1987), detail of a spherically-symmetric local volume fraction in a porous cluster can be obtained via the definition:

$$\phi(r) = \frac{1}{A}\int_A \rho(\mathbf{r})\,dA \tag{2.16}$$

with **r** measured from the centre of mass and A, the area of a spherical shell at radius r. Within this expression, the number density ρ(**r**) has the value unity within a particle and zero in the surrounding fluid. By recognising that the total number of particles in the cluster complies with the expression:

$$\frac{4}{3}\pi a^3 N = \int_0^R 4\pi r^2 \phi(r)\,dr \tag{2.17}$$

Combination of Equations (2.14) and (2.17) via their respective derivatives (dN/dR) leads to:

$$\phi(r) = \phi_0 \left(\frac{r}{a}\right)^{D-3} \quad \text{with} \quad \phi_0 = \frac{A'D}{3} \qquad 0 < r \le R \tag{2.18}$$

Equation (2.18) is often used in theoretical analyses dealing with floc behaviour. While this is useful, it should be noted that in practice flocs are generally heterogeneous (see Figure 3.4) and do not comply with Equation (2.18). In such cases, it is more appropriate to use Equation (2.15). The length scale of a cluster is sometimes characterised by the radius of gyration R_G, this representing the root mean square distance of a particle from its centre of mass (taken as the centre of the cluster). For a spherically-symmetric cluster, the radius of gyration is defined by:

$$R_G^2 = \frac{\int_0^R r^2\, 4\pi r^2 \phi(r)\,dr}{\int_0^R 4\pi r^2\, \phi(r)\,dr} \tag{2.19}$$

When ϕ (r) is defined by Equation (2.18), integration shows:

$$R_G^2 = \left(\frac{D}{D+2}\right) R^2 \tag{2.20}$$

thus providing the correspondence between the radius of gyration and the cluster radius.

Floc material properties are generally a function of the solids volume concentration. At the simplest level, the volume average values of porosity (ε_R) and solids concentration (c_R) are given by:

$$\varepsilon_R = 1 - \phi_R = 1 - A'\left(\frac{R}{a}\right)^{D-3} \tag{2.21}$$

$$c_R = \rho_s A'\left(\frac{R}{a}\right)^{D-3} \tag{2.22}$$

Treating the cluster as a two phase system containing solids and water, the total mass (m_m) and volume (V_m) are specified by $m_m = m_s + m_w$ and $V_m = V_s + V_w$ in which the pairings (m_s,V_s) and (m_w,V_w) identify the mass and volume of the component phases. From this, one may show that solids concentration (m_s/V_m) is defined by:

$$c = \frac{\rho_e}{1 - \rho_w / \rho_s} \tag{2.23}$$

In Equation (2.23), $\rho_e = \rho_m - \rho_w$ refers to the *effective density* of the mixture relative to the suspending fluid (water), the terms ρ_m, ρ_s, ρ_w referring to the density of the mixture, solids and water respectively. By applying Equation (2.23) to a spherical cluster, it follows that its volume average effective density is represented by:

$$\rho_e = A'(\rho_s - \rho_w)\left(\frac{R}{a}\right)^{D-3} \tag{2.24}$$

Comparison with Equation (2.21) shows:

$$1 - \varepsilon_R = \frac{\rho_m - \rho_w}{\rho_s - \rho_w} \tag{2.25}$$

2.1.3.2 *Packing factor*

It is often forgotten that the packing factor A′ is just as important as D in terms of specifying structure and its status needs to be recognised. In the context of Equation (2.14), the value of parameters A′ and D can be gained from the intercept and slope of the log N versus log R plot. This was one approach adopted by Sorensen and Roberts (1997) in conjunction with simulation models for creating particle clusters. Sorensen and Roberts examined the behaviour of the pre-factor (A′) in different topological dimensions for small clusters ($N < 100$). They found that the pre-factor "displays uniform trends with the fractal dimension" *i.e.* it is not an independent variable. For the three-dimensional clusters under scrutiny (resembling those formed under Brownian conditions) they found $D = 1.82$ and a pre-factor of 1.19 (based on R_G as the scaling dimension). Translating this into the form of Equation (2.15) using Equation (2.20) indicates $A' = 0.61$ showing $A' \sim 1$ for small clusters.

In the context of porous aggregates (Section 2.2.2) Gmachowski (1996) shows that the pre-factor corresponds to the term $(r/R)^{1/D}$ in which r is the hydrodynamic radius of the aggregate. Exploiting this approach, Gmachowski was able to simulate some critical hydrodynamic properties of a swarm of porous aggregates using impermeable equivalents.

When applying these concepts to heterogeneous suspensions or to flocs with a multifractal structure, it can be difficult to assign values to parameters to the base parameters. In these situations, one must ensure that either the solids mass or the solids volume fraction within the floc is correctly represented. Here it is helpful to examine Equation (2.22) in terms of the more general expression:

$$c = A'\rho_s\left(\frac{\delta}{\delta_0}\right)^{D-3} \tag{2.26}$$

in which δ and δ_0 refer to the overall size and reference length scale respectively. For simplification we assume that all particles have the same density (ρ_s). It should be stressed that the fractal behaviour is embodied in the identity $c \propto \delta^{D-3}$. In mathematical terms δ_0 and A′ can take arbitrary values, provided they are mutually consistent. Here they lose their physical meaning, and merely represent fitting coefficients which must be consistent with the combination of c, δ and D. Two examples are considered to illustrate some physical restrictions that might be imposed on the definitions of δ_0 and A′ in the context of mass conservation.

Example 1: Mixed sizes

For spherical particles within a spherical cluster of diameter d, the volume average mass concentration is specified by:

$$c = (\rho_s / d^3)\sum_{i=1}^{N} n_i d_i^{\,3} \tag{2.27}$$

The volume average diameter *i.e.* $<d_i^3>^{1/3}$ is defined by the identity:

$$< d_i^3 > = \frac{1}{N}\sum n_i d_i^3 \quad \text{with} \quad \mathrm{N} = \sum \mathrm{n_i} = A'\left(\frac{d}{\delta_0}\right)^D \tag{2.28}$$

Equation (2.27) takes the form of Equation (2.26) with $\delta_0 = <d_i^3>^{1/3}$ and $d \equiv \delta$. Parameter A′ can be adjusted to ensure that c has the correct value. Where there is flow through a porous structure, δ_0 (as an effective size) can be specified via the permeability (Section 2.2.2); discussed in Gmachowski (1998) and illustrated in Section 3.3.7.4.

Example 2: Multifractal structure

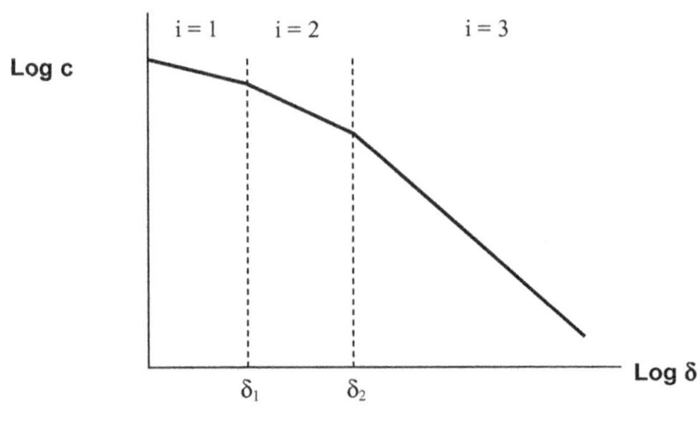

Figure 2.3. Schematic trend in solids concentration-size dependence for an aggregate with a multilevel fractal structure.

In this example, the primary particles are treated as identical, but form part of an idealised structure illustrated in Figure 2.3. In each subregion, it is assumed that the floc is characterised by $c = A_i'(\delta/\delta_0)^{D_i - 3}$ in which i = 1,2,3. Continuity of c across the sub-regional boundaries implies that A_3' is defined by:

$$A_3^{'} = A_1^{'}\left(\frac{\delta_1}{\delta_0}\right)^{D_1-D_2}\left(\frac{\delta_2}{\delta_0}\right)^{D_2-D_3} \tag{2.29}$$

The example illustrates that the pre-factor takes on values that reflect the mass dependencies associated with substructures which may be present. Given that size ranges can spread over several orders of magnitude, it should be noted that the pre-factor (Region 3) can be vastly different from the value (~1) cited in Sorensen and Roberts (1997) for small monofractal clusters.

2.1.4 Chain formation

One of the curiosities of floc growth is the tendency (in the early stages) to form particle chains. This aspect was reviewed in Sonntag and Strenge (1987, p138-140), the central ideas being set out below.

In a development of earlier published work, Rees (1951) made use of DLVO theory (Section 1.2.1) to calculate lines of equal potential around a two-particle aggregate, as a function of the electrolyte concentration. From this, it was shown that the height of the energy barrier at the ends of the particle aggregate was lower than that along the sides – thus favouring growth at the ends. However, the probability of contact depends on the area of contact as well as the height of the barrier. Clearly, as the chain length increases through particle additions, the side area increases, but the two end areas remain constant. At a critical chain length L, the probabilities of contact (including *both* the barrier height and the contact area) at the sides and the ends become virtually equal, and the formation of chains is no longer favoured. This aspect has been closely scrutinised in both theory and experiment and appears to hold quite widely. The application of DLVO theory over a range of conditions indicates values of the ratio L/2a (*i.e.* the number of particles of radius 'a' in the chain) in the range 2 to 12.

Looking at the significance of these findings in the broader context of coagulation structures, Sonntag and Strenge comment "...that the structure of a gel from the coagulation of primary spherical particles is a tangled network of particle chains…". This picture fits well into the characterisation of gels discussed in Section 2.3.3.

2.1.5 Bound water

There is abundant evidence to show that structurally modified water exists at many (or most) water/solid interfaces. Structural effects can show themselves over distances of several hundred molecular diameters (Drost-Hansen, 1977), but one is normally concerned with much smaller length scales. The first few nm away from a surface coincides with the solvation or hydration layers noted in Section 1.2.6. When particle surfaces approach one another *e.g.* during dewatering or under the action of attractive forces, solvent has to be displaced from the intervening space. This becomes increasingly difficult as the spacing is reduced. This is reflected in the magnitude of the bond energy required to release water molecules or in the magnitude of the applied pressure required to overcome the restraining pressures. These are believed to be associated with a combination of dielectric phenomena and influences caused by the finite size of counterions (volume exclusion effect) in proximity of the surface (Paunov *et al.*, 1996). Whatever the cause, the net effect is that water is bound to the surface as a result of force fields which decrease with distance from the surface.

Within the bound water layer (sometimes referred to as vicinal or interfacial water) the modified structure alters its thermodynamic properties (relative to bulk water away from the surface). This affects intrinsic properties, such as freezing point and viscosity. Measurements of such properties provide insight into the extent of the layer affected by structural effects – this being commonly referred to as *bound water*.

For present purposes, we simply acknowledge that bound water will always be present when hydrolysing metal salts are used as a coagulant – thus affecting aspects of floc structure and the dewatering of associated sludges. Once identified, the bound water (BW) is often expressed in the form:

$$BW = \frac{\text{mass of bound water}\,(\mathrm{m_{bw}})}{\text{mass of dry solids}\,(\mathrm{m_s})} \tag{2.30}$$

When bound water is present, it limits the maximum effective density of a floc *i.e.*

$$\rho_e = \frac{\rho_s - \rho_w}{1 + BW(\rho_s / \rho_w)} \tag{2.31}$$

where ρ_w is the density of water. For an individual particle of radius ‘a’, the combined radius (r) of the particle and its encapsulating bound water layer is given by:

$$\frac{r}{a} = \left(1 + BW\frac{\rho_s}{\rho_w}\right)^{1/3} \tag{2.32}$$

There exist many schemes for classifying 'bound water'. Here we follow the scheme described above and refer to the remaining water fraction as 'free water' (in accord with the approach adopted in Sato *et al.*, 1982). Under this description, the floc or sludge mass (m_m) is expressed by Equations (2.33) and (2.34) with m_s, m_{fw} and m_{bw} referring to the component masses of the solids, free water and bound water respectively and θ defining the moisture content.

Equations (2.33) and (2.34) combine to yield Equation (2.35). In terms of Figure 2.4, the term (1 - θ) corresponds to the vertical axis, and the ratio m_{bw}/m_s to the horizontal axis. The fitted trend is obtained when m_{fw}/m_{bw} = 2.8; this means that much of the water remaining after mechanical dewatering must be classified as 'free water', suggesting that the mechanical systems employed did not penetrate the domain of the so-called 'bound water'. It raises the question of the value of such classification schemes, an aspect which is discussed further in Section 3.3.7.2.

$$m_m = m_s + m_{fw} + m_{bw}\,, \qquad m_s = (1-\theta)m_m \tag{2.33, 2.34}$$

$$1-\theta = \left[1 + \frac{m_{fw}}{m_s} + \frac{m_{bw}}{m_s}\right]^{-1} \tag{2.35}$$

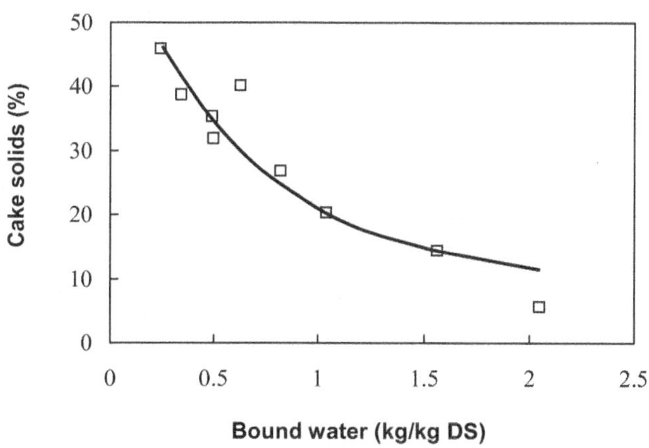

Figure 2.4. Bound water content of alum sludges in terms of the cake solids fraction achieved by mechanical dewatering (based on data reported in Robinson and Knocke, 1992). The trend line is based on Equation (2.35) with m_{fw}/m_{bw} = 2.8 (Bache *et al.*, 1999a).

2.2 HYDRODYNAMIC BEHAVIOUR OF SUSPENSIONS

2.2.1 Low Reynolds number flows

Many situations are encountered in which one is dealing with flow through packed beds, examples including clarification processes, filtration, dewatering or flow through porous flocs. When flows are steady, there is a balance between the pressure gradient driving the flow and the opposing frictional forces. At low Reynolds number, this can be described by Darcy's equation:

$$\frac{dP}{dz} = -\frac{\mu}{k}u \tag{2.36}$$

in which μ is the dynamic viscosity of the fluid, k the permeability and u the superficial velocity. In essence, the permeability (dimension length squared) scales the effective area of the pore space through which the liquid flow takes place.

The *superficial velocity* is defined by Q/A where Q refers to the flow rate through the bed and A its cross-sectional area. The average pore velocity is higher than the superficial velocity because of the volume occupied by the solids and specified by:

$$u_f = \frac{u}{\varepsilon} \tag{2.37}$$

where ε is the volume average porosity of the medium. In Happel's model (*e.g.* Happel and Brenner (1965) or Probstein, 1994), the porous medium is taken as a random assemblage of *cells* each of which contains a particle surrounded by a fluid envelope. The fluid envelope is assumed to contain the same volumetric proportion of fluid to solid as in the entire assemblage. Noting that the pressure drop stems from the dissipation of energy, it follows that drag force per particle (F) divided by the cell volume (V_{cell}) must equal the pressure drop per unit length of bed (both with units N m^{-3}), this allowing Equation (2.36) to be written in the form:

$$\frac{F}{V_{cell}} = \frac{\mu u}{k} \tag{2.38}$$

Noting that the total bed volume can be expressed by:

$$V = \sum V_{cell} = \frac{\sum_i (4/3)\pi r_i^3}{\phi} \tag{2.39}$$

leads to a general expression for the permeability (Gmachowski , 2002a):

$$k = \frac{4\pi \mu u}{3\phi} \cdot \frac{\sum_i r_i^3}{\sum_i F_i} \tag{2.40}$$

where r_i refers to a particle radius and ϕ to the solids volume fraction in the assemblage. The analysis demonstrates that permeability is determined by the prevailing hydrodynamics, as distinct from the physical dimensions of a pore.

The status of the permeability as a control parameter is illustrated by its role in hindered settlement. Following the description provided in Gmachowski (2002a), the superficial velocity is defined by:

$$u = \varepsilon(u_f - u_s) \tag{2.41}$$

in which u_s is the average velocity of the solids. Because the net flow across any cross-section is zero, it follows that liquid flux and the solids flux balance *i.e.*

$$\varepsilon u_f = -(1-\varepsilon)u_s \tag{2.42}$$

The net gravitational force (F_{gn}) on a particle of volume V is $(\rho_s - \rho_{fl})Vg$ in which ρ_{fl} is the fluid density. Referring this to a reference cell shows $V = (1\text{-}\varepsilon)V_{cell}$. Recognising that the ratio F_{gn}/V_{cell} matches the macroscopic pressure gradient, the latter takes the form:

$$\frac{dP}{dz} = (1-\varepsilon)(\rho_s - \rho_{fl})g \tag{2.43}$$

In uniform motion, viscous forces and gravitational forces are in balance. By combining Equations (2.36) and (2.43) and making use of Equations (2.41) and (2.42), it follows that:

$$u_s = \frac{\phi(\rho_s - \rho_{fl})g}{\mu}.k \tag{2.44}$$

This illustrates the impact of both ϕ and k on the settling behaviour of a swarm of particles.

2.2.2 Permeability

There exist many different models for representing the permeability of a particle suspension. Historically the best known of these is the Kozeny-Carman model (*e.g.* Probstein, 1994). Currently, the cell models of Brinkman (1947a) and Happel (1958) have gained favour because they provide a better representation of the flow regime around individual particles. If one considers an isolated sphere of radius R in laminar creeping flow, Stokes law can be expressed in the form:

$$F = f_T u \tag{2.45}$$

in which $f_T = 6\pi\mu R$ is known as a *translational friction coefficient* and u refers to either the streaming velocity or the translation velocity. In a multiparticle system the effective friction coefficient may alter as a result of hydrodynamic interactions between the particles. Addressing the case of a porous spherical cluster of radius R, in which the component particles are uniformly distributed, Brinkman (1947b) expressed the friction coefficient in the form:

$$f_T = 6\pi\mu R\, f(K) \tag{2.46}$$

with

$$K = \frac{k_R}{R^2} \tag{2.47}$$

In this expression, k_R is the internal permeability of the porous sphere and K is dimensionless. The specification of permeability will be considered in due course. At this juncture, it is useful to examine the wider problem of estimating the permeability arising from a swarm of porous spherical clusters. This scenario was examined in Gmachowski (1996) who generated a framework which can be applied to an array of problems involving porous fractal clusters.

According to the Brinkman cell model, the forces experienced by a particle within a cell are affected by both the internal permeability of the individual aggregates (k_R) and the permeability (k) of the total collection of aggregates. The general solution of the equations which control the hydrodynamics of the can be stated in the form of the functional dependence:

$$f(K,\phi_F,\frac{k}{R^2})=0 \tag{2.48}$$

in which ϕ_F is the volume fraction occupied by the aggregates. A solids volume balance across the system shows:

$$\phi_F=\frac{\phi_0}{\phi_R} \tag{2.49}$$

in which ϕ_o and ϕ_R refer to the solids fraction of the primary particles within the system and ϕ_R the average solids fraction within the spherical aggregate. Equation (2.48) defines a family of curves in which K, ϕ_F and k/R^2 appear as base variables.

In order to simplify this system (*i.e.* by reducing the 'family of curves' to a single distribution), Gmachowski (1996) devised a scheme in which a swarm of porous flocs could be represented by an equivalent swarm of impermeable spheres, each being characterised by a hydrodynamic radius r. The transformation between the systems involves the correspondence between R and r, and also necessitates an adjustment in the effective solid volume fraction. An important feature of Gmachowski's approach is that it recognises fractal scaling in the base clusters.

A starting point in the Gmachowski scheme is provided by the mass-radius dependence:

$$N=S_f\left(\frac{R}{a}\right)^D \tag{2.50}$$

in which S_f is a structure factor (a pre-factor) and N the number of primary particles of radius 'a' within the aggregate. The hydrodynamic radius r was defined as the "...radius of an impermeable sphere of the same mass having the same dynamic properties...". The hydrodynamic radius is expected to be proportional to the aggregate radius (R) and must converge to the primary particle radius for the aggregation number when N = 1. These conditions are satisfied by the definition:

$$r=aN^{1/D} \tag{2.51}$$

which leads to the first transformation dependence *i.e.*

$$r = S_f^{1/D} R \tag{2.52}$$

Following permeability matching, Gmachowski (2000) found that r/R could be represented by the empirical function:

$$\frac{r}{R} = \sqrt{1.56 - \left(1.728 - \frac{D}{2}\right)^2} - 0.228 \tag{2.53}$$

From this, it is evident that the structure factor (S_f) shown in Equation (2.50) depends on D, this being illustrated in Figure 2.5. Here it must be emphasised that the base Equations (2.52) and (2.53) are associated with *porous spheres*. In other situations involving isolated *impermeable* flocs, r ≅ R (*e.g.* allowing for departures from sphericity), the fractal dimension *is not constrained* by the assumptions implicit in Equation (2.53).

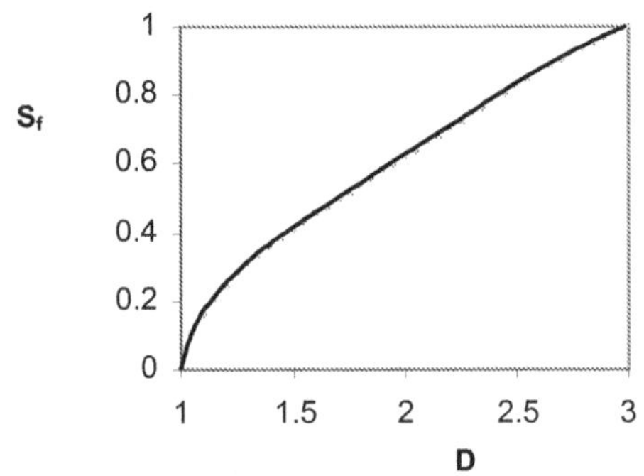

Figure 2.5. Dependence of the structure factor on the fractal dimension.

The replacement of the clusters by impermeable particles of radius r affects the associated volume fraction. If ϕ denotes the volume fraction for the hard spheres, then a comparison of the definitions of ϕ and ϕ_F shows:

$$\phi = \left(\frac{r}{R}\right)^3 \phi_F \tag{2.54}$$

Replacing r/R by use of Equation (2.52) leads to the second transformation dependence:

$$\phi = S_f^{3/D}\phi_F \tag{2.55}$$

When R and ϕ_F are replaced by r and ϕ (via Equations (2.52) and (2.54) with the support of Equation 2.53) and noting that $k_R = 0$ in a hard sphere (implying that K = 0 by Equation 2.47), the solution represented by Equation (2.48) reduces to a single curve represented by:

$$f\left(\frac{k}{R^2},\phi\right) = 0 \tag{2.56}$$

Its functional dependence is described by:

$$U = \frac{5(b+h)^2 + (2b+j)(2b+l)}{20b(b+h)/\phi + 2b(2b+j)} \tag{2.57}$$

in which:

$$U = \frac{9}{2}\phi\frac{k}{r^2} \tag{2.57a}$$

$$b = \frac{3\phi^{1/6}}{U^{1/2}} + 1, \qquad h = -\frac{3}{\sqrt{2U\phi}} - \frac{3}{\phi^{2/3}},$$

$$j = -\frac{4U}{9\phi^2}\left[\frac{27}{4U}\left(\frac{2\phi}{U}\right)^{1/2} + \frac{27\phi^{1/3}}{U} + \frac{135\phi^{1/6}}{(2U)^{1/2}} + 45\right],$$

$$l = -\frac{(2\phi)^{1/3}}{U^{1/2}}$$

A plot of U versus ϕ is shown in Figure 2.6. As well as conveying detail of the permeability, the term U is in the form of a dimensionless settling velocity. Noting that the Stokes formula for an isolated particle is:

$$u_0 = \frac{2}{9}\frac{ga^2(\rho_s - \rho_{fl})}{\mu} \tag{2.58}$$

this combines with Equations (2.44) and (2.57a) to show:

$$U = \frac{u_s}{u_0} \tag{2.59}$$

In Gmachowski (1996), it is emphasised that Equation (2.57) is not confined to monodisperse suspensions formed from spherical particles, but also applies to heterogeneous suspensions formed from irregular particles and even to polymer solutions with an undefined particle size. Gmachowski (2002a) illustrates how a variety of problems involving porous aggregates can be tackled using the equivalent hydrodynamic radius approach.

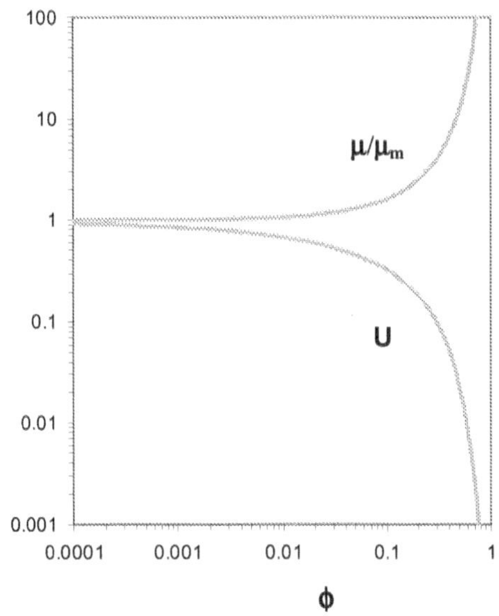

Figure 2.6. Relative viscosity (Equation 2.111) and permeability factor in terms of the volume fraction occupied by impermeable aggregates (based on Gmachowski, 2002a).

2.2.3 Settlement

2.2.3.1 Framework

Here we are concerned with the settlement of individual flocs, an approach which is often used for gaining insight into fractal structure. There are many reviews of this subject *e.g.* Clift *et al.* (1978), Bushell *et al.* (2002) and Tang and Raper (2002).

In steady state, there exists a force balance between the drag force and the net gravitational force this being written in the form:

$$F_{gn} = Vg\,\Delta\rho \tag{2.60}$$

in which V is the envelope volume and $\Delta\rho$ the volume average density difference between the settling object and the suspending fluid. The drag force exerted in a flow with streaming velocity u is represented by the expression:

$$F_D = \frac{1}{2} C_D \rho_{fl} A_c u^2 \tag{2.61}$$

in which C_D is the drag coefficient, ρ_{fl} the fluid density, A_c the projected cross-sectional area perpendicular to the direction of flow. The drag coefficient is sensitive to a multitude of factors such as Reynolds number, shape, surface roughness and internal flows when the object is porous.

If Equations (2.60) and (2.61) are applied to a solid sphere of density ρ_s and diameter d immersed in fluid, the force balance reduces to the expression:

$$C_D \operatorname{Re}^2 = \frac{4gd^2(\rho_s - \rho_{fl})}{3\nu^2 \rho_{fl}} \tag{2.62}$$

in which the Reynolds number is defined by Re = ud/ν. Many different expressions exist for the drag coefficient for a uniform sphere *e.g.* (Fuchs, 1964):

$$C_D = \frac{24}{\operatorname{Re}} + \frac{4}{\operatorname{Re}^{1/3}} \tag{2.63}$$

When Re < 0.5 corresponding to laminar creeping flow, the second term is negligible, and the expression $C_D = 24/\operatorname{Re}$ is equivalent to Stokes law shown by Equation (2.45).

In the development of a framework for describing the effects of shape and porosity, we shall confine the analysis to a description of the kinetics in conditions in which *Stokes law* is applicable. In these circumstances the drag can be represented by a modified form of Equation (2.61) and stated as:

$$F_D = 3\pi d \mu \Omega f_S u \tag{2.64}$$

in which $\Omega \equiv f(K)$ shown in Equation (2.46) (*i.e.* accounting for internal flows in a porous aggregate) and f_S is a shape factor accounting for departures from sphericity: NB, $f_S = 1$ for a smooth sphere. The formal definition of Ω is:

$$\Omega = \frac{\text{Resistance experienced by a permeable sphere of diameter } d}{\text{Resistance experienced by an impermeable sphere of diameter } d} = \frac{f_T}{f_{T0}} \quad \text{...(2.65)}$$

In this expression the term f_T is the translational frictional coefficient defined below Equation (2.45) and f_{T0} its value when $\Omega = 1$. With these inclusions, the drag coefficient for a sphere may be stated in the form:

$$C_D = \frac{24}{\text{Re}} f_S \Omega \quad (2.66)$$

In consequence, the settling velocity takes the general form:

$$u = \frac{1}{f_s \Omega} \cdot \frac{g d^2 \Delta\rho}{18\mu} \quad (2.67)$$

Noting that that the term $gd^2\Delta\rho/18\mu$ defines the terminal velocity (u_o) of an impermeable sphere, it is evident that:

$$\Omega = \frac{u_o}{u} \quad (2.68)$$

It will be seen later that $\Omega \leq 1$ implying that $u \geq u_o$ *i.e.* a porous spherical cluster of particles settles faster than its impermeable equivalent.

2.2.3.2 Shape factors

A wide-ranging review of the theory and methods for dealing with non-sphericity is described in Clift *et al.* (1978), this section drawing on its findings. Much of the theoretical work has been developed for particles with a regular shape and Clift points out that there is really no fully satisfactory method for specifying the drag on irregular particles.

In the context of creeping flows, departures from sphericity imply that the translational resistance terms (roughly equivalent to 'f_T/μ'; see Clift *et al.* (1978, p71) for further detail) gain a directional dependence. The net effect is to alter the effective value of the resistance. The total drag is obtained from a vector addition of the drag associated with each of the velocity components parallel to each of the principal axes of translation. Provided the total drag vector is vertical and passes though the centre of mass, the particle will fall steadily. If the latter condition is not met the particle will rotate and fall in a tumbling motion.

Addressing the case of spheroids, these possess two principal axes of symmetry (axial and normal). Clift *et al.* (1978, p75) summarised the findings of several base studies which describe the dependence of the axial and normal resistances on aspect ratio b/a denoted by E (see Figure 2.7). When these are translated into the component resistances for the case $E < 1$, they are found to be less than $3\pi d$ ($\equiv f_T/\mu$ for a sphere), implying that $f_s < 1$ whereas $f_s > 1$ when $E > 1$. In order to apply these factors to an individual settling particle, one must know its orientation. Expressions exist to generate mean values of the resistance (and thereby f_s) over a set in which the orientation is random.

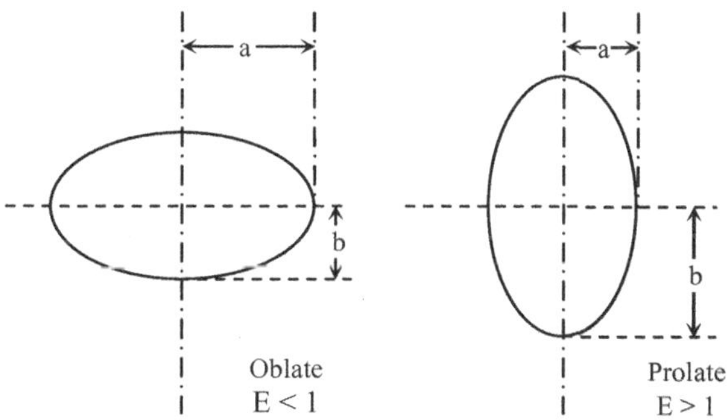

Figure 2.7. Dimensions of spheroids (from Clift *et al.*, 1978.)

In the context of creeping flows, Hill and Power (1956) showed that the Stokes drag on an arbitrary particle "…is less than or equal to that on a body which encloses it and greater than that or equal to a body contained within it…". Thus judicious choice of circumscribed and inscribed *spheres* can give close bounds on the resistance or settling velocity. NB, this approach was advocated in Bache *et al.* (1995) as a suitable approach for estimating R when one has knowledge of the maximum and minimum lengths in an image of a floc cross-section. Because the reference "body" in Hill and Power's statement is interpreted as a sphere, the associated shape factor is unity.

2.2.3.3 Flow-through effects

For flow-through in the Stokes regime, Ω can be determined using the Brinkman model *e.g.* Neale *et al.* (1973):

$$\Omega = \frac{2\beta^2[1-(\tanh\beta)/\beta]}{2\beta^2+3[1-(\tanh\beta)/\beta]} \tag{2.69}$$

where:

$$\beta = \frac{R}{\sqrt{k}} \tag{2.70}$$

Estimates of the permeability may be based on Equation (2.57a) with r interpreted as the [impermeable] primary particle radius. A basic limitation of Equation (2.69) and similar approaches for estimating k, is that they are based on the assumption of uniform distribution of solids.

2.2.3.4 Settling of fractal aggregates

Equation (2.62) (or Equation (2.67) when Re < 1) may be applied to settling aggregates, provided care is exercised in the interpretation of the component terms. For the case of fractal aggregates, the density difference is identified by Equation (2.24). In the case of the porosity factor, there are various routes forward. Numerical modelling by Veerapaneni and Wiesner (1996) on aggregates with radially varying permeability indicated that reasonable estimates of the fluid drag force on a fractal aggregate can be obtained using the volume average porosity shown by Equation (2.21). Where flocs have a multifractal structure, their effective permeability can be estimated using the approach described in Woodfield and Bickert (2001).

An alternative approach is to use the framework described in Gmachowski (2002b, 2003), which is based on a hydrodynamic radius approach. This was developed for embryonic aggregates composed of "less than a dozen or so" particles linked together to form structures ranging from linear chains to compact. The model is based on the calculation of permeability in an aggregate bounded by a circumscribed sphere of known radius (R).

For a settling aggregate (in the Stokes regime) containing n primary particles of radius a, the force balance is expressed by:

$$\frac{4}{3}\pi a^3 n(\rho_s - \rho_{fl})g = 6\pi r\mu u_n \tag{2.71}$$

in which r is the hydrodynamic radius of the cluster and u_n the cluster settling velocity. In this context, r is defined as the "...radius of an impermeable sphere of the same mass, having the same hydrodynamic properties as the real permeable aggregate of radius R...". From this, the relative falling speed (compared with a single primary particle) is given by:

$$\frac{u_n}{u_1} = \frac{n}{(r/a)} \Rightarrow \frac{r}{R} = \frac{n}{(R/a)(u_n/u_1)} \tag{2.72}$$

Within the circumscribed sphere, the packing fraction is given by:

$$\phi = \frac{(4/3)\pi a^3 n}{(4/3)\pi R^3} = \frac{n}{(R/a)^3} \tag{2.73}$$

With knowledge of ϕ, Gmachowski (2002b) recommends the calculation of term β (for inclusion in Equation (2.69)) on the basis of the formula:

$$\beta = \frac{R}{a}\left[\frac{9}{2}\phi\left(1 + \frac{25\phi^{4/3}}{3(1-\phi)^3}\right)\right]^{1/2} \tag{2.74}$$

Crucially, it is recognised that $\Omega \equiv r/R$ (follows from definition of r; see also Gmachowski, 2003). Since r/R is known (Equation 2.69), this value can be substituted into Equation (2.72) to evaluate u_n/u_1. Knowledge of r/R also provides insight into the effective fractal dimension of the aggregate, through use of Equation (2.53).

2.3 RHEOLOGY

2.3.1 General concepts: patterns of deformation

Rheology is the study of deformation and the flow of matter. The most straightforward rheological behaviour is exhibited by Newtonian viscous fluids and by Hookean elastic solids. The dominating feature of colloidal suspensions is their non-Newtonian nature, a symptom being their shear-dependent viscosity behaviour. The origin of such behaviour stems from particle-particle interactions and particle-dispersion medium interactions together with factors such as particle size, concentration and shape and the viscosity of the medium itself (Shaw, 1992).

Because flocs and suspensions are in a continuous state of motion within a treatment works one is always concerned with the rheology. When hydrolysing coagulants are used, the rheology is complicated by the changing nature of the materials which are present as hydrolysis progresses (Section 3.2.7.2).

Here the principal concern is to describe the factors that control the strength of particle networks. Strength factors affect all treatment processes. Before considering this aspect, it is necessary to provide some preliminary grounding.

Valuable foundation work in this field can be attributed to studies such as Michaels and Bolger (1962), van den Temple (1961), Papenhuijzen (1972) and Firth and Hunter (1976a).

In the study of deformation and flow, one is very much concerned with changes in the state of free energy that accompany these processes. For example, energy is required to stretch bonds prior to breakage, this being released when the bonds relax; this component of energy is tied to colloidal forces such as described in Section 1.2. When there is a flow of the liquid medium across the particles, energy is lost (as heat) as a result of the work done in overcoming the viscous forces. One can make a rough distinction between these energy exchange mechanisms, by subdividing the mechanical properties into two groups. The first group covers the response of the structure to small deformations (creep) prior to bond breakage, whereas the second group (following bond breakage) focuses on the flow properties. Before getting involved in detail, it is useful to examine some basic patterns of behaviour.

Figure 2.8 illustrates patterns of deformation in a block of fluid under the action of an external force that acts tangentially per unit area to promote a shear stress σ (Pa). The constant stress is considered to be applied at time t_0 and released at time t_E. Figure 2.8(a) shows the response for an ideal viscous fluid (Newtonian), the shear rate $\dot{\gamma}$ complies with the expression $\sigma = \mu\dot{\gamma}$ in which μ is the dynamic viscosity.

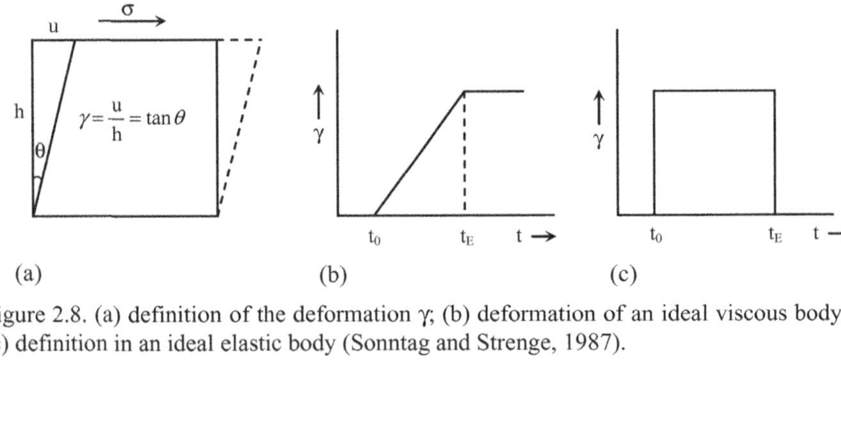

Figure 2.8. (a) definition of the deformation γ; (b) deformation of an ideal viscous body; (c) definition in an ideal elastic body (Sonntag and Strenge, 1987).

When σ and μ are constant, the rate of deformation must be constant, implying that γ increases linearly with time. All the energy invested in the system in the time interval ($t_E - t_0$) is dissipated as thermal energy. When the applied force is released (at time t_E), the state of deformation remains unchanged (Figure 2.8b). In the case of an ideal elastic body (Figure 2.8c), there is a quasi-spontaneous deformation γ_o which remains constant until t_E. On release of the load, the deformation returns spontaneously to its original state. In practice, most particulate suspensions show behaviour in between these ideal types *i.e.* they are *viscoelastic*.

Before discussing the viscoelastic system, it should be noted that in an ideal elastic body, the shear stress and the deformation comply with the relationship:

$$\sigma = G\gamma \tag{2.75}$$

in which G is the shear modulus; this is analogous to Hooke's law in which σ = Ee with σ denoting the tensile stress, E is Young's modulus and e the relative change in length *i.e.* strain.

Focusing on a viscoelastic body (see Figure 2.9), the pure elastic component manifests in two phases. In the first phase, there is the rapid deformation γ_o. Thereafter, there is a time-dependent elastic deformation (γ_R). The viscous component is given by a straight line with slope $1/\mu$. When the load is released (the recovery phase), the elastic deformation components return to zero (their original state). However the viscous part remains at the value ($= \gamma_{total} - \gamma_o - \gamma_R$).

The time dependence γ(t) for this pattern of behaviour in a viscoelastic system may be stated as:

$$\gamma(t) = \frac{\sigma}{G_0} + \frac{\sigma}{G_1}\left[1 - \exp\left(-\frac{t}{\tau}\right)\right] + \frac{\sigma t}{\mu_0} \equiv \gamma_0 + \gamma_1 + \gamma_{cv} \tag{2.76}$$

In the first term, G_0 is the modulus of fast elastic deformation (in Pa). The term G_1 represents the modulus of time-dependent elastic deformation (in Pa), this phase being characterised by a relaxation time (τ). The final term (applying at all times) represents the creep deformation, with μ_o as the creep viscosity.

2.3.2 Viscosity: basic concepts

When flow takes place, work has to be expended against the various force fields which promote cohesion within the fluid medium. Just as bonds are broken, new ones are created when the constituents are in sufficient proximity for cohesive bonds to reform.

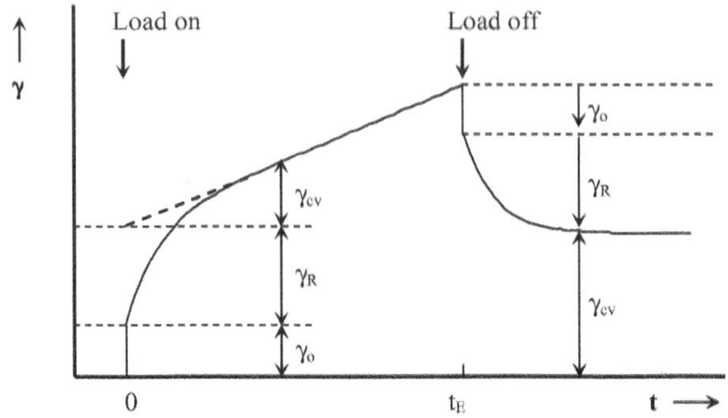

Figure 2.9. Creep behaviour in a viscoelastic body. Symbols comply with Equation (2.76) (adapted from Sonntag and Strenge, 1987).

In the steady state, these processes are in balance. If non-interacting particles are present, additional work or energy is expended in overcoming the drag forces associated with the flow fields which envelope the particles. Generally, the latter aspect comes under the heading of a 'hydrodynamic effect', this being closely tied to concepts discussed in Section 2.2.2. The net effect of these factors is to alter the effective viscosity of the suspension. In a Newtonian fluid, the shear stress tensor is a linear function of the rate of strain tensor. Translating this to a simple shear flow, this can be expressed in the form:

$$\sigma = \mu \dot{\gamma} \tag{2.77}$$

in which σ is the shear stress and $\dot{\gamma}$ is the strain rate (*e.g.* = du/dy). Under this definition, it is seen that μ is a material property expressing the ratio of shear stress to shear rate. At a deeper level, the product $\sigma\dot{\gamma} = \mu\dot{\gamma}^2$ expresses the rate of energy dissipation per unit volume. Particle suspensions exhibit non-Newtonian behaviour, this showing itself in different ways. In the simplest cases, there are 'time-independent' fluids in which the shear stress is a non-linear and single valued function of the shear rate.

A more complex situation arises in a 'time-dependent' fluid in which the shear stress is not a single-valued function of the strain rate, but depends on the shear stress history of the fluid. In the case of viscoelastic fluids, the combination of both viscous and elastic properties causes the shear stress to depend on both the strain rate and the strain. The latter is tied to the elastic character, which introduces a time-dependent behaviour or 'memory' in such fluids. This can show itself as hysteresis.

Focussing on structured suspensions at concentrations below that for dense packing, one finds that the effective viscosity follows the type of pattern illustrated in Figure 2.10. At small stresses or strain rates, a high viscosity (μ_o) results, the flow being Newtonian in character, this corresponding to the creep viscosity noted in Equation (2.76). At very high stress or shear rate, the viscosity again reaches a shear-independent value (μ_∞). The behaviour in this region is in accord with models based on non-interacting particles.

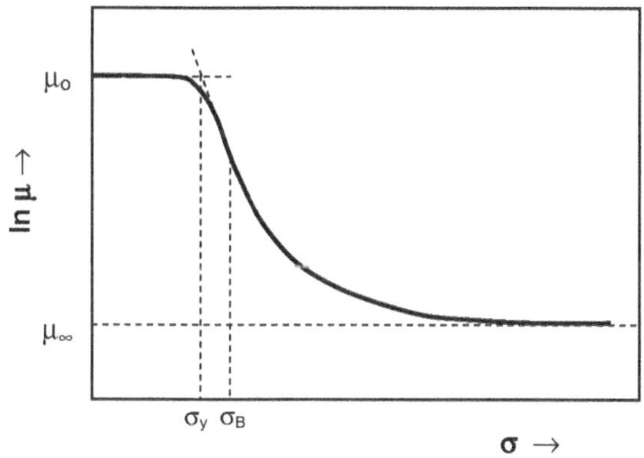

Figure 2.10. General form of a flow curve for a structured dispersion (based on Sonntag and Strenge, 1987).

Between these extremes, the viscosity is strongly shear dependent (*i.e.* non-Newtonian). In this region one may define the yield stresses σ_y and σ_B. The former is termed the *true yield stress*, this being associated with the onset of flow i.e. the transition from a solid-like state to a fluid-like state. In terms of Figure 2.10, Sonntag and Strenge (1987) indicate that the Bingham yield stress corresponds to the inflection point in the μ (σ) plot. A clearer identification of σ_B is gained by a back-extrapolation of the linear region at high shear stress in a plot of σ vs $\dot{\gamma}$ as shown in Figure 2.11; the intercept value at $\dot{\gamma} = 0$ defines σ_B.

The Bingham yield stress corresponds to a condition in which all the bonds that are likely to be broken by the highest stress values have been broken. The remaining clusters correspond to those that will have formed under the highest stress levels *previously encountered* during the aggregation history *e.g.* during rapid mixing. Where particles are weakly bonded, the final state may take the form of primary particles, together with those that are combined temporarily as a result of collisions.

A Bingham plastic is an idealised model represented by the equations:

$$\begin{aligned} \sigma &= \sigma_B + \mu_p \dot{\gamma} \qquad & \sigma > \sigma_B \\ \sigma &= 0 \qquad & \sigma < \sigma_B \end{aligned} \tag{2.78}$$

in which μ_p (corresponding to μ_∞) is known as the *plastic viscosity*. Readers are referred to Buscall *et al.* (1986) for insight into the relative values of σ_y and σ_B for a model suspension. The demarcation between solid-like and fluid-like behaviour is also indicated by the *gel point*.

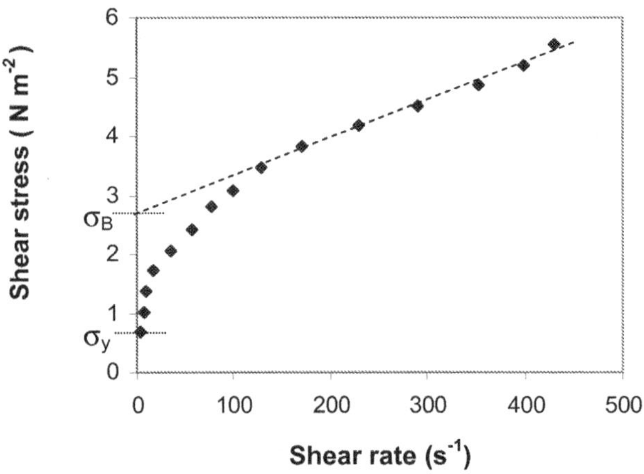

Figure 2.11. 'Equilibrium' flow curve for a 1.5% suspension of coagulated latex showing true and Bingham yield stress (adapted from Buscall *et al.*, 1986).

2.3.3 Gels

A characteristic feature of gels is the existence of a three dimensional coherent network which fills space (see Figure 2.12c). Gelation arises in various ways, for example as a result of overcrowding, whereby individual clusters grow into each other and form bonds between the clusters. This state is indicated by the condition $\phi_F \rightarrow 1$ (see Equation 2.49), implying that the critical solids volume concentration (ϕ^*) $\rightarrow \phi_R$. From this, the critical radius of overlap (R) can be evaluated from the relationship (Piau *et al.*, 1999):

$$\phi^* \approx \left(\frac{R}{a}\right)^{D-3} \tag{2.79}$$

If $\phi^* \cong \phi_g$ (gel point), it suggests that $\phi_g \propto \phi_R$ (Buscall *et al.*, 1988). For submicron spherical particles, Buscall *et al.* (1988) report $\phi_g \approx 0.05$. Many values of this order are reported in the literature. Prior to the formation of relatively rigid bonds associated with gelation, there exists a stage of transient gel formation which occurs at lower volume fractions *e.g.* < 0.001 (Tanaka *et al.*, 2005).

When a gel is formed, small angle and neutron scattering analysis *e.g.* des Cloizeaux and Jannink (1987), indicates that a network may be characterised by a mesh size ξ - the size scale being roughly equivalent to the scale of the mesh as measured by the distance between nodes (where three or four chains join).

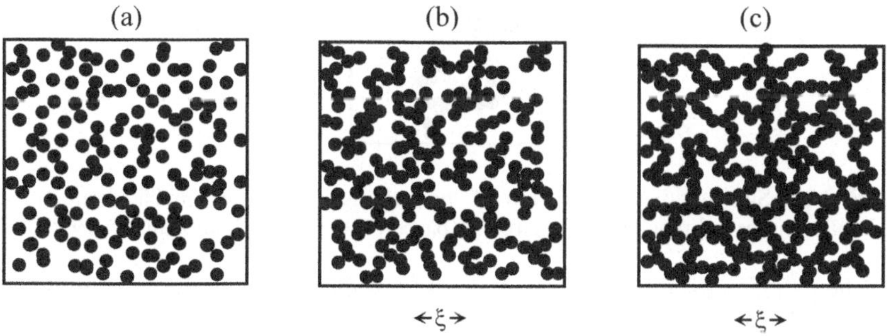

Figure 2.12. Schematic representation of gel formation during fractal aggregation. (a) early stage; (b) state of almost-contacting aggregates; (c) gel state (adapted from Uriev and Ladyzhinsky, 1996).

More strictly, ξ is the characteristic radius of the particle density correlation function and is often referred to as the *correlation length*. The relation between the mesh size and the number of particles $N(\xi)$ within the mesh is specified by $N(\xi)/\xi^3$. Treating this as a fractal aggregate of radius ξ, leads to:

$$\xi \cong a\phi^{-1/(3-D)} \tag{2.80}$$

$$\xi \cong R\left(\frac{\phi}{\phi^*}\right)^{-1/(3-D)} \tag{2.81}$$

$$N(\xi) \cong \phi^{-D/(3-D)} \tag{2.82}$$

In these relationships, the packing factor is assumed to be ≈ 1.

When addressing the load-bearing behaviour of a structure, it is recognised that branches, such as illustrated in Figure 2.13, do not transmit load - the load transmission takes place along the so-called *backbone*. In terms of elastic behaviour and particularly deformation, a distinction is made between regions of the backbone in which particles are connected in series to form a 'single bonding' pathway as distinct from parts of the backbone with double or multiple bonding pathways (schematically connected in parallel). In terms of deformation the latter offer greater stiffness and lack the flexibility of single bonds.

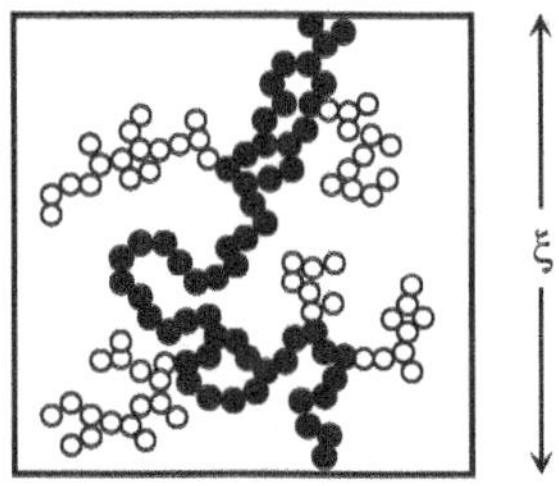

Figure 2.13. Schematic representation of the gel mesh: particles on backbone (black circles); particles in branches (open circles) (based on Uriev and Ladyzhinsky, 1996).

When the focus is on the most flexible components of the backbone, the number of particles found in single bonding pathways in a mesh of size ξ is estimated by:

$$N_s \approx \left(\frac{\xi}{a}\right)^{D_s} \tag{2.83}$$

where D_s is the fractal dimension of the single bonds. Values of D_s are not known for aggregates of different types, but can be estimated from the behaviour of the analysis of elastic shear modulus.

2.3.4 Fast elastic deformation

When a structure of the type illustrated in Figure 2.12(c) is deformed without rupture under the action of a small shear stress (σ) at constant composition, volume, temperature and pressure, the change in the free enthalpy (per unit volume) at the deformation γ_0 is given by:

$$\left(\frac{\partial G_0}{\partial \gamma}\right) = \left(\frac{\partial U}{\partial \gamma}\right) - T\left(\frac{\partial S}{\partial \gamma}\right) = \sigma \tag{2.84}$$

in which U is the internal energy (equal to the interaction energy) and S the entropy. Under the conditions cited $G_0 = \sigma/\gamma_o$ where γ_o is the deformation. There have been very differing approaches for evaluating the theoretical form of G_0, which is seen to involve both the entropic and non-entropic components. Though some detail is provided in Sonntag and Strenge (1987), the approach does not take account of a fractal structure. Here, it is advantageous to outline the arguments of Uriev and Ladyzhinsky (1996) and Piau *et al.* (1999).

Uriev and Ladyzhinsky considered that the modulus of a gel is of the order of its mesh modulus, *i.e.* of a fractal aggregate of size ξ. The load is transmitted from mesh to mesh via its backbone as distinct from branches (see Figure 2.13), which do not transmit load. The configuration of the backbone is tortuous and its projections in all directions attain the order of ξ. When the mesh is subjected to a force F, it transmits a stress on the order of F/ξ^2. Under the action of force the backbone also bends. Multiple pathways (*e.g.* parallel strands, such as illustrated in top half Figure 2.13) were regarded as regions of stiffness in which bending could be neglected *i.e.* the deformation is distributed amongst the single bonds. If $\delta\theta$ is the change in angle between pairs of particles connected by single bonds, the total backbone bending angle ($\equiv \gamma_0$) is then approximately $N_s\delta\theta$.

Under this description:

$$G_0(\xi) \cong \frac{F/\xi^2}{N_s \delta\theta} \tag{2.85}$$

Following an estimate of $\delta\theta$, Uriev and Ladyzhinsky showed that the modulus is specified by:

$$G_0 \cong kG_p\phi^\alpha \qquad \alpha = \frac{(3+D_s)}{(3-D)} \qquad \phi > \phi_g \tag{2.86}$$

in which G_p is the modulus of the particle material, k a coefficient whose value is in the range 0.0001 to 1 and D_s defined by Equation (2.83). Values of α are shown in Table 2.1.

Rather than using the force picture, Piau *et al.* (1999) invoked an energy argument – assigning a constant free energy value E_f to domains of size ξ; E_f corresponds to the stress conditions at which flow is *just* initiated and is a measure of the work (see below) that is expended in reaching that state. The authors argue that the critical deformation (γ_c) for this condition can only involve two length scales *i.e.* ξ and R (the critical radius of overlap) implying that $\gamma_c = \xi/R$ (defined by Equation 2.81). The elasticity modulus is then calculated by:

$$G_0 \approx \frac{E_f}{\xi^3} \cdot \frac{1}{\gamma_c^2} \qquad (2.87)$$

from which they found that $G_0 \propto \phi^{5/(3-D)}$. Though this differs from Equation (2.86), the deductions are broadly equivalent.

Steady flows only reveal the viscous properties of a suspension. For this reason, valuable information about the viscoelastic properties is gained by subjecting a suspension to an oscillating stress in which the amplitude and frequency of the oscillation (ω) are altered. This procedure provides insight into the dynamic shear modulus $G^*(\omega)$. One finds that the stress and strain are out-of-phase. In these situations, it is mathematically expedient to represent G^* as a complex function *i.e.* $G^*(\omega) = G'(\omega) + iG''(\omega)$ where $i = \sqrt{(-1)}$. The real part of this function, $G'(\omega)$, is called the *storage modulus* (*i.e.* referring to stored energy) and gives the in-phase stress to strain ratio. The *loss modulus* (*i.e.* for dissipated energy) $G''(\omega)$ represents the out-of-phase stress to strain ratio. The magnitude of G_0 is given by the modulus $|G^*(\omega)| = (G'^2+G''^2)^{1/2}$ at low frequency. In practice this is not very different from G' ($\omega \rightarrow 0$) (Piau *et al.*, 1999). For further information on oscillatory rheometry see Marin (1988).

Table 2.1. Values of exponents appearing in Equations (2.86), (2.108), (2.93) and (2.109) for a set of D and D_s reported in Uriev and Ladyzhinsky (1999).

D	D_s	α	β	δ	ζ
		$G_0 \propto \phi^\alpha$	$\tau_s \propto \phi^\beta$	$\sigma_y \propto \phi^\delta$	$\mu_0 \propto \phi^\zeta$
1.8	1.3	3.6	1.1	2.5	4.7
1.8	1.6	3.8	1.4	2.5	5.2
2	1.3	4.3	1.3	3.0	5.6
2	1.6	4.6	1.6	3.0	6.2
2.2	1.3	5.4	1.6	3.8	7.0
2.2	1.6	5.8	2.0	3.8	7.8

2.3.5 Yield stress in shear

When one addresses the question of yield, it is really a question that centres on a phase transition from a solid-like state to a fluid-like state. If one inspects Figure 2.12(c), this provides an intuitive view of the issues involved.

The gel state is identified by a continuity of connections or bonds between the component particles or subunits. This is somewhat like a fishing net in which the knots (nodes) are linked to each other by strands (analogous to a backbone of characteristic size, ζ). If most of the connecting strands (of whatever size) are broken, then one has the type of situation illustrated in Figure 2.12(a) in which isolated clusters exist *i.e.* do not form part of a continuous network. If a fraction (P) of the strands is broken, then it is still possible for the net to behave as a continuous structure in spite of the presence of large internal 'holes'. In percolation theory e.g. Stauffer (1979) (which addresses phase transitions), it is recognised that there exists a critical fraction (P_c) such that if $P < P_c$, an 'infinite' network extends (percolates) through the space (*e.g.* Figure 2.12c), whereas when $P > P_c$ only finite clusters remain (*e.g.* Figure 2.12b). Within the domain $P_c < P < 1$, the region $P \cong P_c$ i.e. close to transition state, maps the conditions associated with the true yield stress and is the region in which the viscosity is at its highest value (see Figure 2.10). In Uriev and Ladyzhinsky (1996), the regime $P = 1$ corresponds to a state of "…disconnected space filling aggregates…" and links the yield stress σ_y to the state $P \approx 1$. The Bingham yield stress is identified by plots of the type shown in Figure 2.11. Under conditions of intense shear, aggregates are broken further and one is left with a system of either primary particles or high strength aggregates that can resist the shearing forces - perhaps resembling the regime illustrated in Figure 2.12 (a).

Studies reported in Buscall *et al.* (1986, 1988) on polystyrene latex suspensions indicated that both σ_y and σ_B displayed a similar dependence on the solids volume fraction and particle radius 'a', this being in the form $\sigma_y \propto \phi^\alpha a^\beta$ with $\alpha \approx 3$, $\beta \approx -2$. In the case of σ_B, other studies e.g. Firth (1976) show dependencies described by the values $\alpha = 2$ and $\beta = -1$.

2.3.5.1 'True' yield stress

The 'true' yield stress marks the transition from a solid-like state to the fluid-like state, a regime that is not well understood. In the first instance, it is instructive to consider the 'static' model described in Piau *et al.* (1999) which is based on the same rationale as Equation (2.87). This yields the definition:

$$\sigma_y \approx \frac{E_f}{\xi^3} \frac{1}{\gamma_c} \tag{2.88}$$

in which $\gamma_c = \xi/R$. Making use of Equation (2.81) shows $\sigma_y \propto \phi^{4/(3-D)}$.

Equation (2.88) provides useful insight into the meaning of E_f, noting that the product $\sigma_y\gamma_c$ measures the work done per unit volume by the stresses in reaching the yield state. In other literature *e.g.* Sonntag and Russel (1987), the term E_f/ξ^3 corresponds to the critical value of "the local energy of distortion"; this forms the basis of the Mises yield criterion noted in Bisplinghoff *et al.* (1965) and Adler and Mills (1979).

One may then progress to the 'dynamic' model described in Ladyzhinsky and Uriev (1996). Focussing on the transition state, the authors proposed a model in which regions of solid-like state and fluid-like state coexist within the gel around the yield condition. By assigning proportions of these states across the shear cross-section, the authors were able to offer a possible explanation of the viscosity-shear stress behaviour and characteristics of the shear modulus (in an oscillatory shear mode) in this domain. In terms of the patterns shown in Figure 2.10, their definition of σ_y was based on the condition:

$$\sigma_y = \mu_0 \dot{\gamma}_c \tag{2.89}$$

in which $\dot{\gamma}_c$ is a critical shear rate. Employing the analogy of a reference network (see above), Uriev and Ladyzhinsky (1996) correlated the energy dissipation with the number of network strands that are broken. They assigned E as a quantity of mechanical energy necessary to break a gel connecting structure within a volume ξ^3. On recovery, the connecting structure transforms the energy quantity $\overline{E}$ into heat, this taking place during the timescale t_r. In order for a system to maintain a given P value (*i.e.* from energy input and losses), the average energy dissipation per unit volume must be set by $P\overline{E}/t_r\xi^3$. This implies:

$$\sigma\dot{\gamma} \cong P\overline{E}/t_r\xi^3 \tag{2.90}$$

where σ is the shear stress.

It is assumed that only γ controls the recovery time at such $\dot{\gamma}$ values, the recovery being achieved by adjacent broken meshes coming across one another and re-establishing bonds; this takes place in a time frame set by $t_r \approx 1/\dot{\gamma}$. By assuming that $\sigma_y \equiv \sigma$ at the state P = 1, the conceptual model indicates:

$$\sigma_y \cong \overline{E}/\xi^3 \tag{2.91}$$

The authors note that the quantity, $\overline{E}$, is likely to be closely correlated with the minimum energy needed to break the backbone connecting structure (*i.e.* with the single bond energy E) showing:

$$\overline{E} \cong kE \tag{2.92}$$

in which k(~1) is a coefficient of proportionality. With this information Equation (2.91) takes the form:

$$\sigma_y \cong \frac{kE}{a^3}\phi^\delta \qquad \delta = \frac{3}{3-D} \tag{2.93}$$

Values of δ are listed in Table 2.1. If $E \propto a$ (*e.g.* Equation 1.2), then $\sigma_y \propto a^{-2}$ in accord with the findings of Buscall *et al.* (1988).

2.3.5.2 Bingham yield stress

In contrast to the true yield stress, the Bingham yield value has received far more attention. One of the simplest interpretations of σ_B which permeates the literature of aggregates formed by (weak) depletion forces (*e.g.* Yan *et al.*, 2000) is the model (Prestige and Tadros, 1988) based on the formula:

$$\sigma_B = NE_{sep} \tag{2.94}$$

in which E_{sep} is the separation energy *per contact* and N is the number of contacts between particles in the system. Hence σ_B scales the amount of energy required to separate the floc into single units. In this model, N is defined by:

$$N = \frac{3n\phi}{8\pi a^3} \tag{2.95}$$

in which *n* is the average number of contacts per particle (*coordination number).* According to Torres *et al.* (1991), the coordination number within an aggregate of fractal dimension D, can be expressed by the form:

$$n = B2^D \tag{2.96}$$

in which B is a structure-dependent pre-factor.

When dealing with flowing sol exhibiting plastic behaviour, estimates of σ_B are often based on the approach advocated in Michaels and Bolger (1962). In this regime it is argued that the rate of energy dissipation per unit volume (E_{tot}) can be broken down into a number of separate contributions covering flows around flocs, interactions between flocs, losses within flocs and so on. Some of these contributions are more significant than others. Restricting our discussions to conditions of high shear, analysis reported in Firth and Hunter (1976b) suggested that E_{tot} was dominated by two contributions:

$$E_{tot} = \sigma\dot{\gamma} = E_{DR} + E_V \tag{2.97}$$

in which $E_V = \mu\dot{\gamma}^2$ is the viscous energy dissipated by the suspension medium flowing around the flocs and $E_{DR}(= \sigma_B\dot{\gamma})$ the energy dissipation associated with the rupture of floc doublets. Schematically this is equivalent to the model of a Bingham plastic (see Equation 2.78). Floc doublets are formed when two flocs collide and combine, some energy being release and dissipated as heat. For the breakage of a doublet a certain amount of energy is required, this being supplied by the shear flow. This energy consists of two parts: (i) the energy required to break the bonds between the two flocs forming the doublet and (ii) the energy needed to stretch the bonds within the flocs, as the tension must be transmitted from the shear field to the floc-floc interface. Of these two parts, most of the energy dissipation is associated with the stretching processes rather than the rupture.

Before probing this further, it is necessary to translate these phenomena into the situation found in a continuously stirred flow. In the steady state, there is a balance between floc creation and floc rupture. The rate of flocculation is controlled partly by the combination of the collision frequency and the adhesion coefficient (α; see Section 3.3.1.2). According to Von Smoluchowski (1916), the frequency of collisions per unit volume for flocs of radius 'a' and volume fraction ϕ_F is specified by:

$$3\phi_F^2\dot{\gamma} / \pi^2 a^2 \tag{2.98}$$

If this is multiplied by the trapping efficiency (α), it determines the rate of doublet formation and thereby the rate of destruction. For each doublet broken, the amount of energy released per unit volume can be specified by the total separation energy term (E_{sep}). Combining these processes leads to the identity:

$$\sigma_B = \frac{3\alpha\varphi_F^2}{2\pi^2 a^2} E_{sep} \tag{2.99}$$

This is a much foreshortened account of the energy expenditure and the factors which control σ_B.

There is much debate about the specification of E_{sep} and E *etc.*, because of the variety of contributory components, some being noted above. Both Firth and Hunter (1976b) and van de Ven and Hunter (1977) note that much of the energy loss can be attributed to viscous energy dissipation accompanying liquid movements within the floc during deformation. Equations (2.94) and (2.99) (with the support of Equations (2.49) and (2.15)) provide contrasting views of the dependence of σ_B on the solids volume concentration and neither explain the dependence ($\sigma_B \propto \phi^3$) reported in Buscall *et al.* (1986). The sensitivity of σ_B to the zeta potential was demonstrated in Firth (1976) - this exemplifying the role played by colloidal forces. Clearly, the behaviour of σ_B is system-dependent, and care must be exercised in extrapolating from one regime to another.

Figure 2.14 shows an example of the Bingham yield stress for a polymer-free, 'fresh' sample of alumino-humic sludge gained from a coloured and low-turbidity water. Data were measured using a Brookfield digital viscometer and are considered to be most reliable in the range $c < 10^4$ mg/l. It should be noted that the value c = 20,000 mg/l ($\phi \approx 0.01$) corresponds roughly to the conditions found in a 300 μm floc of the same type; this indicates that such alumino-humic flocs with diameters exceeding 300 μm are likely to have Bingham yield stress values below 0.6 N m^{-2}.

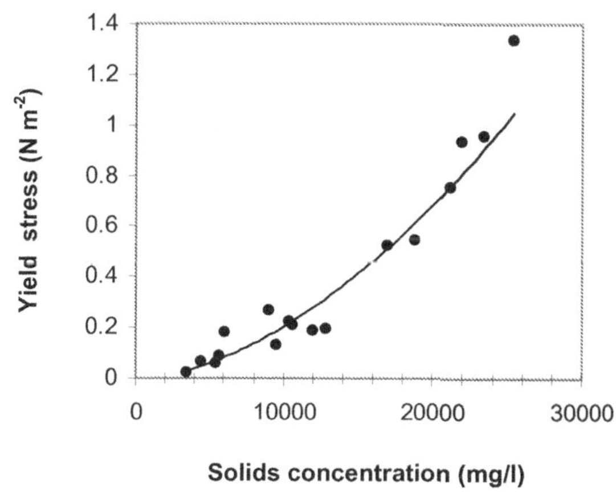

Figure 2.14. Example of dependence of Bingham yield stress on solids concentration for an alumino-humic sludge. The trend line is a power law fit consistent with the variation $\sigma_B \propto \phi^{1.8}$ (unpublished data, DHB).

2.3.6 Normal stresses

Normal stresses are of particular concern in consolidation processes such as thickening, sludge dewatering and centrifuging. They also result from the integrated effect of surface stresses acting on particles and flocs. The appraisal set out in Section 2.4 suggests that floc rupture is often associated with straining forces rather than shear.

2.3.6.1 *Multifractal elasticity*

When a blob of matter is stretched in one dimension by a force F, the tensile stress is defined by the ratio $\sigma = F/A_c$ in which A_c is the cross-sectional area. Considering the case of a rectangular bar of length l, breadth b and height h made of material, which is described by classical elasticity theory, the lateral deformations (thereby affecting A_c and thereby the effective stress) are described by:

$$\frac{\Delta b}{b} = \frac{\Delta h}{h} = -\nu_P \frac{\Delta l}{l} \tag{2.100}$$

in which ν_P is the Poisson ratio ($0 < \nu_P < 0.5$). A constant value of ν_P implies linearity between the lateral deformation and the extension. These concepts have to be modified when the structure is fractal. Balankin (1996) addressed this task in a theory of elasticity applying to random multifractal structures. Some features are described below.

When a force is applied to a blob possessing a multifractal structure, it leads to the appearance of a unique new characteristic scaling length L_F. Under the action of different forces, say F_1 and F_2, the *relative deformation* can be described by the scaling parameter:

$$\lambda_F = \frac{L_{F_1}}{L_{F_2}} \tag{2.101}$$

Lateral deformations in different space dimensions are defined by:

$$\lambda_i = \frac{L_i}{l_i} = \lambda_F^{-\nu_F}, \quad i = 2, 3, \ldots, d \tag{2.102}$$

in which L_i/l_i is the relative change in the lateral scaling lengths and:

$$\nu_F = \frac{D_B}{d-1} - 1 \tag{2.103}$$

In Equation (2.103), ν_F is the transverse deformation component, which coincides with Poisson ratio at infinitely small strains. Otherwise it shows that the lateral deformations are both non-linear, with a dependency on the fractal dimension D_B (box-counting; Section 2.1.2.1) and the corresponding space dimension (d).

Balankin's analysis was founded on the second law of thermodynamics, whereby the force obeys:

$$F = \left(\frac{\partial U}{\partial L_F}\right)_T - T\left(\frac{\partial S}{\partial L_F}\right)_T \tag{2.104}$$

In this expression, the first term describes the energy component of the internal forces and the second term is the entropic component. Expressions were derived for ΔU and ΔS involving D_B, the correlation dimension D_C and the information dimension D_I (tied to the entropic contribution; see Section 2.1.2.2).

When the structure is monofractal $D_B = D_C = D_I$, the corresponding stress (σ_{11}; consistent with i = 1) was shown to be defined by:

$$\sigma_{11} = \frac{E}{1+6\nu_F+4\nu_F^3}\cdot\left[(\lambda_1^{1+4\nu_F}-1)-2\nu_F(\lambda_1^{-1-2\nu_F^2}-1)\right] \tag{2.105}$$

where $E = (\partial\sigma_{11}/\partial e_{11})$ is Young's modulus. At infinitesimally small strains, $|e_{11}| = (|\lambda_1^2-1|)^{1/2} \ll 1$, Equation (2.105) reduces to its classic counterpart $\sigma_{11} = Ee_{11}$ (*i.e.* Hooke's law).

2.3.6.2 Normal yield stress

In the approach described by Buscall and White (1987) for the consolidation of concentrated suspensions, the strength of a network under uniaxial compression was considered to be governed by a combination of the compressive yield stress $P_y(\phi)$ and a permeability term. Permeability is a necessary feature because drainage from the network is a rate-determining feature in the dynamics of collapse. A formal definition of $P_y(\phi)$ is given by:

$$\frac{D\phi}{Dt} = \begin{cases} 0 & P_s < P_y(\phi) \\ \kappa(\phi)[P_s - P_y(\phi)] & P_s > P_y(\phi) \end{cases} \tag{2.106}$$

in which $\kappa(\phi)$ is a rate constant and P_s is the transmitted stress or 'particle pressure'. Various studies *e.g.* Channell and Zukoski (1997) show that $P_y(\phi)$ and $\sigma_y(\phi)$ (the shear yield stress) have a linear relationship, though this is not always the case. Where these parameters have both been measured on sludges of different type, the ratio $P_y(\phi)/\sigma_y(\phi)$ has been found to lie in the approximate range 10 to 100. Various explanations have been proposed *e.g.* in Uriev and Ladyzhinsky (1996), but none have been fully vindicated.

In contrast to the compressive yield stress, there is an acute lack of information concerning the tensile stress of fractal structures. One suspects that this will be of the same order as the shear stress – but this is a guess. In the practical situation, it can be deduced from a force balance approach. For example, Yeung and Pelton (1996) used a micromechanical technique to pull single flocs apart to measure their tensile strength. Their findings were subsequently represented in a different form by Boller and Blaser (1998), as shown in Figure 2.15. This shows the association between the tensile strength and size, the size conveying information about the impact of the solids volume concentration on strength.

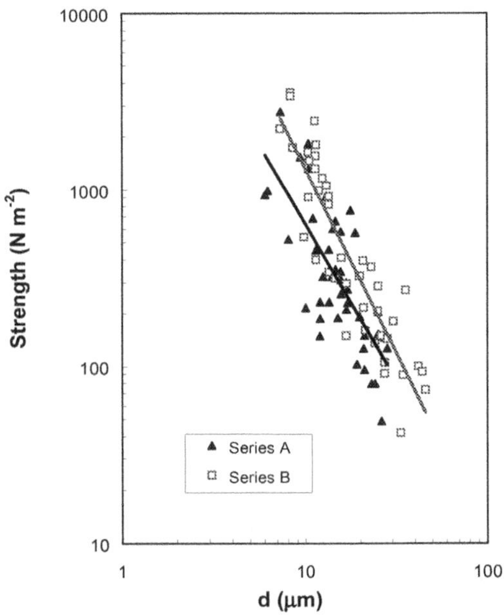

Figure 2.15. Tensile stress-size dependence based on data from Yeung and Pelton (1996). Aggregates were formed from a precipitated calcium carbonate suspension coagulated with water-soluble polymers (termed A and B) (from Bache, 2004).

2.3.7 Viscosity

Figure 2.10 shows the inter-relationship between the viscosity and shear stress, the trend being characterised by the values μ_0 and μ_∞ as $\sigma \rightarrow 0$ or ∞ respectively. Data reported in Uriev and Ladyzinsky (1996) indicate that the change in viscosity between these extremes can be far more abrupt. They note that the magnitude of the creep viscosity can be estimated using:

$$\mu_0 \cong G_0 \tau_s \tag{2.107}$$

where τ_s is the stress relaxation time for the solid-like state; this characterises the timescale associated with the breaking of the backbone. From thermodynamic equilibrium arguments with respect to the stability of mesh elements, Uriev and Ladyzhinsky proposed:

$$\tau_s \cong \tau_0 \left[\exp\left(\frac{E}{kT} \right) \right] \phi^\beta \qquad \beta = \frac{D_s}{(3-D)} \tag{2.108}$$

in which E is the single bond energy and τ_0 is the period of the thermal oscillation of the single bond (~10^{-11} s). Substitution of Equations (2.86) and (2.108) into Equation (2.107) yields:

$$\mu_0 \cong kG_p \tau_0 \left[\exp\left(\frac{E}{kT} \right) \right] \phi^\zeta \qquad \zeta = \frac{3+2D_s}{(3-D)} \tag{2.109}$$

Sets of β and ζ are listed in Table 2.1. For a 2.5% silica gel, Uriev and Ladyzhinsky cite $\tau_s \cong 10^5$ s and $G_0 \approx 10^3$ Pa, yielding $\mu_0 \cong 10^8$ Pa s. When $\sigma > \sigma_y$ the fluid-like viscosity (μ_F) can be estimated using $\mu_F \cong G_0 \tau_F$ in which τ_F is the corresponding relaxation time. In some respects μ_F behaves in the same way as μ_∞ shown in Figure 2.10 in the sense of representing a limiting behaviour. Uriev and Ladyzhinsky cite $\tau_F = 10^{-4}$ s and μ_F (~μ_∞?) $\cong 10^{-1}$ Pa s. Here, the salient features are the major changes in viscosity and relaxation timescales when the yield stress is exceeded. Such timescales have significance when one is evaluating the influence of transitory stress events, such as encountered in a turbulent flow (see Baldyga and Bourne, 1995).

In conditions of high shear stress, the limiting viscosity μ_∞ is analogous to the plastic viscosity (μ_p) shown in Equation (2.78). These terms scale the viscous energy dissipated by the suspension medium flowing round the flocs. In the first instance, the flocs can be treated as impermeable particles. For dilute suspensions of hard spheres ($\phi < 0.02$), Einstein (1906) showed that the viscosity was expressed by:

$$\mu = \mu_m (1 + 2.5\phi) \tag{2.110}$$

in which μ_m is the viscosity of the fluid medium. This does not depend on the size of the spheres or their distribution. The relationship is often applied up to $\phi \sim 0.1$. For similar suspensions at higher concentrations, there exist many formulae, *e.g.* Happel (1958):

$$\frac{\mu}{\mu_m} = 1 + 5.5\phi\left[\frac{4\phi^{7/3} + 10 - (84/11)\phi^{2/3}}{10(1-\phi^{10/3}) - 25\phi(1-\phi^{4/3})}\right] \tag{2.111}$$

A plot of this relationship is shown in Figure 2.6. Gmachowski (1996) showed how this formula could be applied to a swarm of porous spheres by making use of the transformations expressed by Equations (2.52) and (2.55).

2.4 FLOC STRENGTH

2.4.1 Perspective

During its passage through a water works, a particle suspension encounters a wide variety of fluid regimes. In processes such as flocculators, or during transfer between treatment stages, the flow will be fully turbulent with flocs subjected to relatively high levels of stress, whereas in settling processes or within the interstices of a filter bed, the stress conditions are lower. At all times, the suspension or deposit will tend to adjust to the prevailing conditions.

In the case of a suspension, the process of adjustment affects the whole floc size distribution, an equilibrium or quasi-equilibrium state representing a balance between growth and rupture (Section 2.5). Although all flocs are subjected to hydrodynamic stress, attention tends to focus on the largest flocs, because these are the weakest. Many studies have shown that their size follows a relationship of the form:

$$d_{\max} = \frac{C}{G^m} \tag{2.112}$$

in which d_{max} is the diameter of the largest flocs, G the velocity gradient. Parameters m (>0) and C are fitting coefficients, the latter being interpreted as a 'strength constant' on the grounds that G is a property of the flow field.

In terms of deposits, one is also concerned with matters of strength because of their susceptibility to scour. In many respects, flocs and deposits are similar in that they are both particulate gels with a fractal structure. However, they encounter differing stress regimes. The external forces acting on a deposit are dominated by viscous stresses because of the existence of a thin boundary layer adjacent to the surface within which the flow is laminar. This arises because the velocity at a non-slip surface is zero. In contrast, flocs held in suspension are subject to a wider range of flow-related forces - particularly when the flow is turbulent.

Unlike surface tension, the strength of agglomerates is not a material property, but depends on the cohesive forces between particles, their solids volume concentration and way that the solids are structured. Only in the very simplest of situations can one predict the forces and their net influence on strength with any degree of confidence. More often, information about their strength is gained by subjecting flocs to known shear conditions or by direct measurement as in Yeung and Pelton (1996; Figure 2.15). By analysing the conditions in which rupture takes place, estimates can be gained of the critical yield stresses *e.g.* Blaser (2000a, 2000b). For a mechanical failure to occur, it is necessary that a component of the applied stress should exceed some critical yield value. However the exceedence of a yield stress does not constitute a sufficient condition for estimating strength, because it does not take account of the time factors involved. In turbulent fields, forces are transitory and deformation itself is not instantaneous (see Figure 2.9).

An alternative to the 'force-balance' approach is to examine the critical state from the viewpoint of energy *i.e.* to determine the work that needs to be expended against the binding forces and viscosity to achieve separation. Examples of this approach were described in Section 2.3.5. The advantage of the energy approach is that it provides a more accurate view of the physical processes involved in rupture. Because energy is a scalar, the energy approach circumvents difficulties arising from the vector nature of forces. For example, one does not need to distinguish between normal and shear stresses or their corresponding yield values within the object under stress.

In this section, the aim is to provide insight into the dynamics of the rupture process and to show how this links with Equation (2.112). This is rather like a complicated jigsaw in which there are many pieces of information that must be put in place before the overall picture can emerge. A useful starting point is to consider some of the issues in the context of viscous flows.

2.4.2 Viscous flows

2.4.2.1 Surface stress

When the Reynolds number is small, the hydrodynamic stress acting on a surface complies with the scaling:

$$\sigma \propto \mu E \tag{2.113}$$

where E is the strain rate ($\equiv \dot{\gamma}$). In viscous flows, E is specified by the velocity gradient (units: s^{-1}). Across a wide flat surface $|\boldsymbol{\sigma}/\mu E|$ will be spatially uniform, whereas across an object immersed in the flow, the stress vector $\boldsymbol{\sigma}$ will vary in magnitude and direction according to its position and the shape of the surface, in addition to the shape, $\boldsymbol{\sigma}$ will also depend on the structure of the enveloping flow.

2.4.2.2 Flow fields

Figure (2.16) shows examples of flow fields which are often used in studies of floc strength.

The most familiar of these is the simple shear flow (Figure 2.16a) in which the velocity in the x direction is defined by u = Ey. When a floc is encompassed by this flow, the velocity gradient induces rotation and causes stretching along the x-axis. Shear flows can be generated in a Couette apparatus.

A two-dimensional convergent flow of the type shown in Figure 2.16(b) can be generated using a Taylor four-roll mill (see Blaser, 2000b). In this flow, the streamlines are hyperbolic, the velocity components u and v in the x and y directions being expressed by u = Ex and v = -Ey. As in the case of a simple shear flow, the velocity gradient terms e.g. du/dx are uniquely determined by E. A floc following a streamline will be compressed in the y direction and stretched in the x direction.

Figure 2.16(c) illustrates a three-dimensional contraction flow. This is convergent in the xy plane and extends along the z-axis. Schematically, this type of pattern arises when flow enters the pore of a filter bed. High velocity gradients can arise in contraction flows, a feature which might be responsible for floc breakage. Salient features of a contraction flow are described in Blaser (2000a) and Boller and Blaser (1998) and are outlined below.

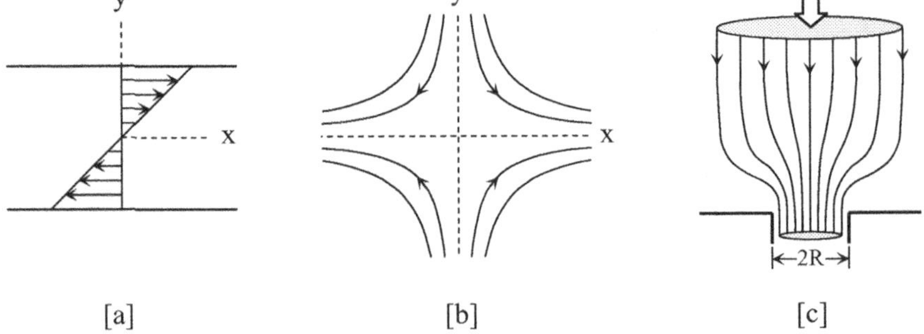

Figure 2.16. Illustrative flow fields: (a) two-dimensional shear flow, (b) two-dimensional convergent flow, (c) three-dimensional contraction flow.

2.4.2.3 Induced stresses

Bagster and Tomi (1974) examined the stresses on planes within a sphere in a number of simple flow fields with a view to predicting the manner of break-up under fluid forces. The emphasis of their study was to gain a better appreciation of the stresses within a sphere. From knowledge of the flow field around a sphere, the stresses acting on its surface were established. This information, together with knowledge of inertial forces induced by spin, was used to calculate the net force acting on an arbitrary dividing plane that splits the sphere into two portions.

The dividing plane can take any orientation with respect to the incident flow; in essence it represents a potential failure surface. The integrated effect of force components normal to the dividing plane (of known area) results in an average tensile stress acting on the dividing plane, whereas the sum of force components parallel to the dividing plane provides information about the average shear stress. In the case of a two-dimensional shear flow (Figure 2.16a), Bagster and Tomi showed that the maximum normal stress within the sphere was 5 μE, this acting on a dividing plane through the sphere centre and oriented at 45° to the incident flow. Thus, if the tensile strength of the material in the sphere is less than 5 μE, a failure will occur. The maximum shear stress (8μE) in a simple shear flow acts on a dividing plane through the sphere centre and facing the incident flow. Corresponding analysis was carried out by Jeffrey (1982) to assess the tension acting across the midplane of a sphere in an axisymmetric straining flow. In this case, the flow under consideration was a uniaxial extension, with the flow converging towards the z-axis in the xy-plane and extending along the ± z axis (Jeffrey; private communication). Data cited in the paper indicates the normal stress acting on the mid-plane was 37 μE. This is very different from the 5 μE for a sphere in a shear flow and emphasises the sensitivity of the stress regime to the properties of the enveloping flow.

The case of the contraction flow shown in Figure 2.16(c) is of interest. For a pore of radius R, a velocity scale is specified by the definition $u_0 = Q/2\pi R^2$ with Q as the flow rate. From theory described in Blaser (2000a), the axial velocity w (through the centre-line of the pore) is defined by:

$$w = \frac{Q}{2\pi(z^2 + R^2)} \quad \text{or} \quad \frac{w}{u_0} = \frac{1}{1 + (z/R)^2} \tag{2.114}$$

From this, the axial velocity gradient is specified by:

$$\frac{\partial w}{\partial z} = -\frac{Q}{2\pi}\frac{2z}{(z^2+R^2)^2} \quad \text{or} \quad \frac{R}{u_0}\frac{\partial w}{\partial z} = -\frac{2(z/R)}{(1+(z/R)^2)^2} \tag{2.115}$$

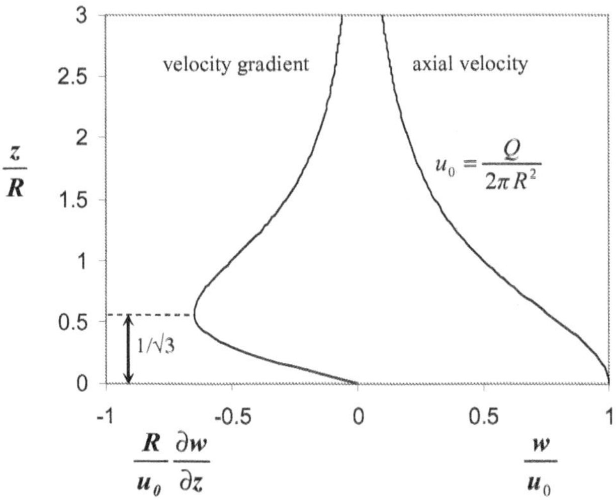

Figure 2.17. Axial velocity and velocity gradient in a contraction flow above a porous filter bed (Boller and Blaser, 1998).

A plot of these functions is shown in Figure 2.17. This shows the existence of a maximum in the velocity gradient at the vertical position z/R= $1/\sqrt{3}$. This has value:

$$\frac{R}{u_0}\frac{\partial w}{\partial z} = -\frac{3\sqrt{3}}{8} \tag{2.116}$$

For pore diameters and filtration rates used in practice, Boller and Blaser (1998) indicate that the maximum velocity gradients above sand filters is likely to be in the range the range 5 to 300 s^{-1}. If one applies these findings to the case of a spherical particle or floc of diameter d, which follows the axial streamline, it will experience a maximum strain force specified by:

$$F_{max} = \frac{5}{4}\pi d^2 \mu E = \frac{15\sqrt{3}}{64}.\frac{Q\mu d^2}{R^3} \tag{2.117}$$

Blaser (2000b) examined the implications of shape by considering the pattern of stress acting over the surface of an ellipsoid (with semi-axes a, b and c) in a convergent strain field of the type illustrated in Figure 2.16 (b). The surface stresses were reduced to a single resultant force (f_r) acting over half the ellipsoid's surface (see Figure 2.18). Data was presented in the form of a three-parameter plot involving $f_r/\mu c^2E$ together with the aspect ratios b/c and a/c. In the case of a sphere (a = b = c), it was shown $f_r/\mu c^2E = 5\pi$ which contrasts with the situation a/c = b/c ≈ 3 for which $f_r/\mu c^2E > 64$. From an appraisal of the stress pattern over the ellipsoid surface, Boller and Blaser (1998) concluded *that rupture was likely to be caused by strain rather by shear.* Information of this type was used to estimate the strength of a latex suspension coagulated with ferric chloride using sweep floc as the destabilisation mode. Flocs with a characteristic size in the range 30-400 µm were found to have an average strength specified by:

$$\left\langle \frac{2f_r}{S} \right\rangle = (8 \pm 3).10^{-2} \quad \text{N m}^{-2} \tag{2.118}$$

with S as the surface area of the ellipsoid (Blaser, 2000a).

Bagster and Tomi (1974) also analysed the surface stresses acting on a sphere under Stokesian settlement. They showed that the maximum shear stress acting on surface points on the horizontal mid-plane perimeter was $3u\mu/2a$ where u is the settling velocity and 'a' the sphere radius. This can constrain the size of a floc during settlement. For example when there is a large concentration of flocs with individuals settling at different speeds, there are continuous opportunities for growth as a result of inter-floc collisions. When flocs coalesce, they are likely to have greater water content and diminished strength. At some stage, a condition is reached in which strength and shear stresses come into balance, thus limiting growth.

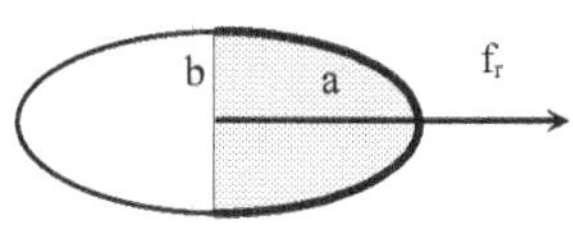

Figure 2.18. Resolved axial force f_r acting on an ellipsoid half surface.

So far our discussion has focussed on solid objects immersed in a flow. When the object is porous, one must also take account of the stresses induced by the internal flow. For example Adler and Mills (1979) examined the rupture behaviour in a porous settling sphere with a view to gaining a better understanding of the factors that control the limiting size of a freely settling cluster. Sonntag and Russel (1987) pursued a similar analysis for a porous spherical cluster with fractal structure in a shear flow.

2.4.3 Turbulent flows

2.4.3.1 Stress fields

Turbulent flows are dominated by vorticity. In the vortices, which are locally distinct (rather like tornados), the flow is rotational whereas between the vortices, the flow is irrotational. The latter has flow structures similar to those described for laminar flows *i.e.* shear fields, stream flows and convergence zones.

However an important difference from the viscous flows shown in Figure 2.16, is the random and temporary nature of the flow patterns associated with turbulence. This implies that one has to approach the problem of specifying stress fields in a different way. From the background information provided in Section 1.4.2, it was evident that turbulent motions have a strong dependence on scale *i.e.* eddies of different size (see Figure 1.17). In terms of rupture, the eddies, which deliver the critical forces acting on a floc are those of similar size. Larger eddies simply move the floc around within the flow, whereas smaller eddies nibble away at the floc surface - perhaps causing local erosion. Because flocs are small (usually less than a few mm), it implies that the critical size scales of interest are those found in the fine scale structure of turbulence. In many respects this is fortunate, because it greatly simplifies the task of expressing the dynamics. Within the fine structure (Section 2.4.3.2), Kolmogorov (1941) argued that the properties of the turbulence are governed by the separation distance (l) between two points. For the reasons stated above, the relevant spacing can be specified by the floc diameter *i.e.* $l \cong d$.

At large Reynolds number, Batchelor (1951) showed that the instantaneous pressure difference between two points A and B in a turbulent flow in a turbulent flow is specified by:

$$\overline{(P_A - P_B)^2} = \rho_{fl}(\overline{(u_A - u_B)^2})^2 \tag{2.119}$$

where u_A and u_B refer to instantaneous values of velocity fluctuations at points A and B and ρ_{fl} refers to the fluid density. The basic interdependence between the pressure fluctuations and the velocity differences also holds at low Reynolds numbers, but includes a proportionality coefficient (~1), which depends on a turbulent Reynolds number. From this, the average normal stresses can be determined provided the spatial velocity differences are known. This aspect is discussed in the next section. In the Kolmogorov theory, the choice of what to include in the right hand side of Equation (2.119) depends on the relative values of l and the Kolmogorov length (η) (Section 1.4.2.1). For spacings $l << \eta$, Equation (2.125) holds, whereas as when $l >> \eta$, Equation (2.124) holds. With this information we may write the turbulent stress in the form:

$$\sigma_t \sim \rho_{fl}(\varepsilon / \nu) l^2 \qquad l << \eta \tag{2.120}$$

$$\sigma_t \sim \rho_{fl}(\varepsilon l)^{2/3} \qquad l >> \eta \tag{2.121}$$

Equation (2.121) is restricted to very high Reynolds number flows, a situation that is unlikely to occur in a waterworks.

In many respects, Equation (2.120) is of greater relevance in spite of the stated size restriction. The term $(\varepsilon/\nu)^{1/2}$ defines the local strain rate in the viscous domain. In this context it is useful to draw attention to remarks of Jeffrey (1982) who stated "...as far as a particle is concerned, it experiences a laminar flow field, whose strength varies in time in a way determined by the statistics of turbulence. The flow field will be a linear function of position and can be described by a rate of strain tensor (**E**) and a rotation vector (**Ω**). The rotation simple causes the particle to spin with the flow and break-up is caused by the straining flow (described by **E**)...". Jeffrey's views fit neatly into the role played by vortex structures in micromixing as described by Baldyga and Bourne (1995) and noted in Section 1.4.2.2. In a turbulent field, the smallest vortex systems which persists, is on the scale of $l/\eta \approx 12$ (Figure 1.18). These structures entrain the surrounding fluid, a process that will tend to both trap and concentrate flocs within the vortex. Once there, they will be subject to the dynamics described by Jeffrey. Observations of breakage events in stratified turbulence (described in Bache *et al.*, 1995) and Bache & Rasool, 1996) have provided examples of flocs 'exploding', rather than just being pulled apart as in a straining field.

When this occurs, it indicates that the floc is enveloped by a flow structure in which the turbulence kinetic energy is very intense. These observations are also consistent with the views of Hinze (1955) - who indicated that turbulent rupture involves large-scale deformations [affecting the floc as a whole]. This differs markedly from the breakage mechanism described in Sonntag and Russel (1987) for a porous floc in which the outer layers of a floc 'peel off' when a critical condition is exceeded.

2.4.3.2 *Fine structure: specification of turbulence energy*

The framework for describing floc rupture depends on knowledge of the fine structure of turbulence - particularly the distribution of energy at different scales. In many respects this centres on the description of the right hand side of Equation (2.119). Although some detail of turbulence energy was provided in Section 1.4.2.1, this did not address the viscous domain.

Kolmogorov (1941) argued that the small scale motions of turbulence (defined by the velocity difference $\Delta u(l)$ over the distance l) were uncorrelated with the large scale and that their statistics, e.g. $\langle(\Delta u)^2\rangle$ would be universal, provided the motions are defined in terms of their *relative velocity*. The statistical features of turbulence are defined by structure functions of form:

$$G_p(l) = \left\langle \Delta u(l)^p \right\rangle = \left\langle (u(x+l) - u(x))^p \right\rangle \tag{2.122}$$

where p represents the order of the structure function. Kolmogorov's first similarity hypothesis states that relative velocity distributions are uniquely determined by ν and $<\varepsilon>$ provided $l << \Lambda$ (Λ as shown in Figure 1.17). Under these conditions, the structure function must be of the form:

$$G_p(l) = \mathrm{v}_K^p \beta_p\left(\frac{<\eta>}{l}\right) \tag{2.123}$$

in which $\beta_p(<\eta>/l)$ is a universal function and v_K is the Kolmogorov velocity scale defined by Equation (1.33). In the foregoing, discussion will be confined to the second order structure function (p = 2), which refers to the kinetic energy per unit mass $<\Delta u(l)^2>$.

In the domain $l >> \eta$ (the inertial range starting at about $l/\eta > 35$) (*e.g.* Herweijer, 1995) Kolmogorov's second similarity criterion requires that structure functions are determined by ε and l alone, *i.e.*

$$< u(l)^2 >= C_2[\langle\varepsilon\rangle\, l]^{2/3} \tag{2.124}$$

in which C_2 is a universal constant (in the range 1.0-2.0). On the basis of theory described in Batchelor (1947) for viscous dominated motion, Bache (2004) demonstrated that $<\Delta u(l)^2>$ was well represented by:

$$< \Delta u(l)^2 > \approx \frac{1}{15}(< \varepsilon > \nu)^{1/2} \frac{l^2}{< \eta >^2} \quad \text{for the range} \quad 0 < \frac{l}{< \eta >} < 3.5 \tag{2.125}$$

Equations (2.124) and (2.125) have strict ranges of applicability and fail to define $<\Delta u(l)^2>$ in the intervening region. This problem can be circumvented through use of Equation (2.123) with β_2 defined empirically (Batchelor, 1951) as shown in Equation (2.126) in which $x = l/<\eta>$ and $C_2 = 2.0$.

$$\beta_2(x) = \frac{x^2/15}{\left[1+(15C_2)^{-3/2}x^2\right]^{2/3}} \tag{2.126}$$

2.4.3.3 Rupture criterion

Rather than basing the rupture criterion on a force balance, the energy approach has many advantages. Rupture involves the supply of kinetic energy to carry out the necessary work against binding and viscous forces to achieve separation. A model of this type was proposed in Bache *et al.* (1999b) and developed further in Bache (2004). In this approach, the rupture criterion was based on the energy statement:

$$\frac{1}{2}\rho_{fl}\mathrm{v}_d^2 \sim \frac{6n_b\varphi}{\pi d^3} \tag{2.127}$$

The left hand side of Criterion (2.127) expresses the locally available turbulence kinetic energy per unit volume on the spacing scale d equivalent to the floc diameter, whereas the right hand side expresses the necessary amount of work ($k\varphi$) that must be carried out within the domain of the floc volume ($\pi d^3/6$) to achieve separation. The term n_b, refers to the number of bonds which are stretched and broken and φ defines a quantity of energy [irrespective of its origin] which is associated with each bond during the separation process.

The framework described in Bache (2004) and noted in Section 2.4.3.2 allows the left hand side of the criterion to be defined with a reasonable degree of precision across all size scales within the fine structure of turbulence. The term v_d^2 is equivalent to the term $\Delta u(l)^2$ when l = d. For example substituting Equation (2.125) into the left hand side of Criterion (2.127) shows:

$$\frac{1}{2}\rho_{fl}\mathrm{v}_d^2 \approx \frac{1}{30}\rho_{fl}(<\varepsilon>\nu)^{1/2}\frac{d^2}{<\eta>^2} \tag{2.128}$$

This indicates that the available kinetic energy per unit volume vanishes as $d \to 0$. It also shows that the available energy depends on the average value of ε at a particular location, and also on the $d/<\eta>$ ratio, where $<\eta>$ is the average

Kolmogorov length in the locality of interest (NB, ε varies with space and time; see Baldyga and Bourne, 1995).

To make further progress with Criterion (2.127), we require more detail of its right hand side. Before this is considered, it is useful to refer to the information shown in Figure 2.15; this illustrates that strength decreases with increasing size. The way in which the left hand side and the right hand side of Criterion (2.127) respond to the floc size is shown schematically in Figure 2.19. When the size scale is small, the strength is relatively high and there is insufficient available kinetic energy to promote breakage. It is only when these factors have similar magnitude that rupture can take place, this determining the critical size.

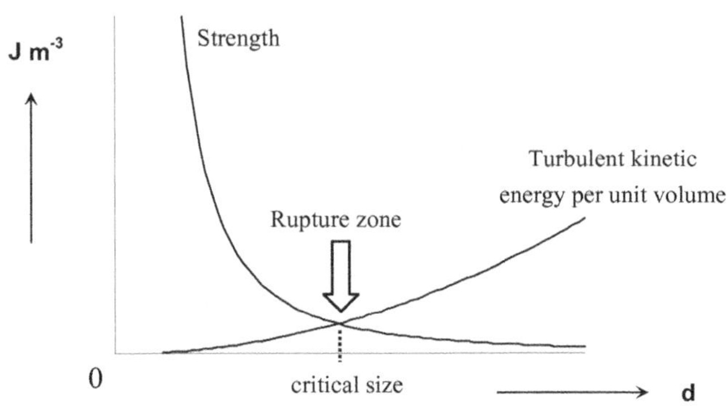

Figure 2.19. Interaction of turbulent kinetic energy and strength leading to definition of critical size.

2.4.4. Structural model

The right hand side of Criterion (2.127) can be interpreted as 'strength' (S) on the grounds that the free energy per unit volume expresses a thermodynamic pressure (see also Section 2.3.5.1). Hence we define:

$$S = \frac{6 n_b \varphi}{\pi d^3} \tag{2.129}$$

In Bache *et al.* (1999b) it was assumed that the number of bonds broken was scaled by the number of particles within the floc (N), *i.e.*

$$n_b = \beta N \tag{2.130}$$

with β as a proportionality coefficient. If the floc is formed from primary particles of diameter d_o, and density ρ_s then a mass balance shows that N is defined by:

$$N = \frac{c\,d^3}{\rho_s d_o^3} \tag{2.131}$$

where c is the solids mass concentration within the floc defined by Equation (2.22). The number of particles in a floc is also be expressed by Equation (2.14). With these inclusions Equation (2.129) takes the form:

$$S = S_o\left(\frac{d}{d_o}\right)^{D-3} \quad \text{where} \quad S_o = \frac{6\beta\varphi A'}{\pi d_o^3} \qquad (2.132), (2.133)$$

in which A′ represents the pre-factor. The term S_o can be interpreted as a strength scaling factor. For the domain D < 3, Equation (2.132) shows that strength diminishes with increasing size, this being consistent with the trend shown in Figure 2.15. Comparison of the structure of Equations (2.312) and (2.24) shows the association between strength and effective density - the common feature being the solids volume concentration. This stems from the assumption implicit in Equation (2.130). Although the structural component controls the strength-size dependence, it is S_o which carries the 'strength' information which characterises the floc and gives it identity. The correspondence $S_o \propto A'$ shows the impact of the packing factor on strength. Similarly, the factor βφ conveys information about the binding energy.

Sonntag and Russell (1987) argued the form of the fractal dependence of strength in a different way. From scrutiny of existing data, they recognised that elastic properties (such as those discussed in Section 2.3) were generally of the form:

$$\sigma = \sigma_0 \phi^r \tag{2.134}$$

with σ_0 and r as coefficients. With the floc solids volume concentration defined by Equation (2.15), Equation (2.134) can be restated as:

$$S = S_o^{'}\left(\frac{d}{d_o}\right)^{r(D-3)} \tag{2.135}$$

with S_o' as a scaling term. It is seen that Equation (2.132) is a particular case (r = 1) of Equation (2.135) with $\sigma_o \equiv S_o/A'$.

2.4.5 Strength-turbulence interaction

Having established the identity of the components appearing in Criterion (2.127), its evaluation is illustrated for the regime $0 < d/<\eta> < 3.5$ through use of Equations (2.128) and (2.135) *i.e.*

$$\frac{1}{30}\rho_{fl}(<\varepsilon>\nu)^{1/2}\frac{d^2}{<\eta>^2} \sim S_o^{'}\left(\frac{d}{d_o}\right)^{r(D-3)} \tag{2.136}$$

For convenience of expression (rather than physical significance), the symbol '~' is dropped and treated as an equality. After some manipulation this can be represented by the expression:

$$\frac{d}{d_o} = \left(\frac{30S_o{}'\nu}{\rho_{fl}d_o^2<\varepsilon>}\right)^{\frac{1}{2+r(3-D)}} \tag{2.137}$$

The equivalent expression for the inertial range $d/<\eta> >> 1$ is:

$$\frac{d}{d_o} = \left(\frac{2S_o{}'}{C_2\rho_{fl}d_o^{2/3}<\varepsilon>^{2/3}}\right)^{\frac{3}{2+3r(3-D)}} \tag{2.138}$$

where $C_2 \approx 2$. A more general framework is described in Bache (2004) which caters for all $d/<\eta>$ ratios. For comparison with Equation (2.112), it is useful to represent Equations (2.137) and (2.138) in the form:

$$d = \frac{\gamma}{<\varepsilon>^{m/2}} \tag{2.139}$$

For example, taking the case of Equation (2.137):

$$\gamma = \left[\frac{30 S_o' \nu \, d_o^{r(3-D)}}{\rho_{fl}} \right]^{\frac{1}{2+r(3-D)}} \quad \text{and} \quad m = \frac{2}{[2+r(3-D)]} \qquad (2.140), (2.141)$$

Equation (2.139) isolates <ε> from the remaining factors which control the floc size. If γ and m remain constant during mixing (generally observed), it should be recognised that their component terms, as shown in Equations (2.134) - (2.135), are insensitive to the mixing conditions. This is useful, because it allows one to gain information about base parameters such as S_o' and r from the values of γ and m.

To establish the linkage between Equations (2.139) and (2.112), it must recognised that <ε> has a strong spatial dependence. Shinnar (1961) showed that $<\varepsilon> \propto \bar{\varepsilon}$ where $\bar{\varepsilon}$ is the volume average value of <ε> specified by Equation (1.39). Hence we may write:

$$<\varepsilon> = f\bar{\varepsilon} \qquad (2.142)$$

in which f denotes a position-dependent coefficient of proportionality. With this identity, the 'strength constant' C appearing in Equation (2.112) is defined by:

$$C = \frac{\gamma}{(f\nu)^{m/2}} = \left[\frac{30 S_o' d_o^{r(3-D)}}{f\rho_{fl}} \right]^{\frac{1}{2+r(3-D)}} \qquad (2.143)$$

This shows that C depends on the structure of the mixing regime (through f) as well as the strength factor S_o'. Hence the practice of interpreting C as a surrogate for strength is misleading.

Table 2.2. Summary of floc strength data in association with Equation (2.132) (from Bache, 2004).

Floc type	d (μm)	d_0 (μm)	D	S_o ($N\ m^{-2}$) No polymer	S_o ($N\ m^{-2}$) Polymer
Al-HA	120-250	0.008	1.24	$(0.27\text{-}0.82)\times10^6$	
Al-humic	100-3000	0.008	1.1	$(2.3\text{-}11.5)\times10^6$	$(30\text{-}558)\times10^6$
Al-kaolin	200-1600	3.5	≈ 1.9	0.10-0.50	6.2-8.1

Table 2.2 summarises information that can be gained from analysis of plots in the form of Equation (2.112) based on data reported in Bache (2004). The data refers to a humic acid (HA) suspension, a suite of low turbidity coloured

waters (humic) and a clay suspension reported in Tambo and Hozumi (1979). In all cases the coagulant was aluminium sulphate, sometimes used in conjunction with a polymer.

The S_o data refers to an inter-quartile range (based on the original observations) and permits the direct calculation of strength. For example: a 200 μm alumino-humic floc without polymer conditioning is adequately represented by the set $S_o = (2.3 - 11.5) \times 10^6$ N m^{-2}, $d/d_o = 25000$ and D = 1.1. Substitution into Equation (2.132) shows strengths in the range (0.01 - 0.05) N m^{-2}. This is less than the Bingham yield value indicated in Section 2.3.5.2 for similar flocs of comparable size.

The primary effect of the polymer is to alter the magnitude of S_o. Making use of Equation (2.137) for identical mixing conditions, the impact of S_o on size is indicated by the ratio:

$$\frac{d_P}{d_{Po}} = \left(\frac{S_o(P)}{S_o(0)}\right)^{m/2} \tag{2.144}$$

where d_{Po} and d_P are respective maximum sizes in the absence and presence of polymer and $S_o(0)$ and $S_o(P)$ the corresponding strength scales. For example when m = 0.5 and strength scales in the ratio 16:1, Equation (2.144) shows that the maximum floc size doubles when the polymer is present; this is not too far from what happens to alumino-humic flocs when subjected to the typical 'floc-strengthening' polymer dosages used in a practice.

2.5 DISTRIBUTIONS

2.5.1 Overview

When suspensions pass through a treatment system it should be recognised that solid liquid separation processes often discriminate on the basis of particle size, or may depend on 'bulk' factors, such as permeability or floc volume concentration. In these situations, knowledge of distributions should represent an essential element in design and the understanding of process performance.

Although there may be variations in flow and raw water quality, one is often dealing with distributions that are in steady state or slowly varying. Distributions alter as the suspension passes through the treatment system, this stemming from changes in the shear regime and of greater consequence, the removal of material in separation processes. Generally there is very little published information on this theme, though aspects were addressed in Lawler (1997). Examples of questions that need to be addressed include: What is the size distribution in a flocculator? How are distributions affected by shear? What is the character of a suspension entering a rapid gravity filter?

Given sufficient time for adjustment, distributions tend to be unimodal (Figure 3.22) or possess an exponential dependence on size (Rosen, 1984). In practice, there may be insufficient time for distributions to attain these forms. Here, the intention is to illustrate a general approach for representing unimodal distributions and to set this in the context of mass conservation. The latter is important for two reasons. First, whatever the shape of the distribution, one must be able to account for the bulk solids concentration across the distribution. Second, solid-liquid separation processes are often judged in terms their impact on the mass load.

Distributions are often represented by an array of characteristic parameters such as the mean size, standard deviation, or percentiles. These parameters are tied to statistical features of the underlying number-size distribution and can be formally defined using distribution moments (Section 2.5.3; see also Table 3.11). As well as providing descriptors of the number-size distribution, moments can yield information about other features of the distribution, such as surface area, volume or the distribution of mass (examples in Section 3.2.7.3). These parameters interface with separation processes in different ways. For example adsorption processes depend on surface area. Volume comes into play in settlement processes (Section 3.3.4) and affects head losses in deep bed filtration (Section 3.3.6). Where separation processes discriminate on the basis of size, the selected cut-off condition (*e.g.* determined by the treatment rate in a clarification process, Section 3.3.1) affects the mass removal efficiency.

Superficially, the impact of shear on the size distribution stems from its influence on the cut-off size (d_{max}), which behaves in accord with Equation (2.112). Beside the responses of d_{max} to shear, the underlying balance between the rate of aggregation and the rate of fragmentation is also altered, the entire distribution adjusting to a new state. Experimental evidence *e.g.* Spicer and Pratsinis (1996a, 1996b) has demonstrated that when G is altered the distribution can be self-preserving. This feature implies that a collection of size distributions $\psi(d)$ can be represented a single shape function $\psi(u)$ in which u is a scaled size, such as $u = d/d_{max}$ (*e.g.* Figure 3.22). When distributions are self-preserving, it becomes easier to describe the properties of distribution in response to changes in the process variables. In statistical terms the equilibrium profile is the most probable outcome of a large sequence of random events constrained by factors such as mass conservation. Drawing on classical statistical mechanics, Rosen (1984) argued that in these circumstances the information entropy is at its maximum - the information entropy being a statistical measure of the unpredictability in a chaotic system (Section 2.1.2.2). Gmachowski (2001) explored this approach as a means of specifying the character of the mass distribution associated with Brownian motion. One outcome of the analysis was to show that the distribution shape and the fractal index D were linked. Distributions can also be estimated by numerical simulation of the underlying processes. In the context of coagulation processes these are usually based on the population balance approach developed by Von Smoluchowski (1917). The principal difficulty with this approach lies in the simulation of the floc kinetics.

In the foregoing, we provide a formal description on a framework for characterisation distributions, illustrating principles using the gamma distribution (Equation 2.151). Examples of distributions and the features of the framework are presented in Section 3.2.7.3. Material appearing in the remainder of Section 2.5 and Section 3.2.7.3 is drawn from Bache and Rasool (2006). As in Gmachowski (2001), flocs are treated as monofractal – a limitation that severely restricts the application of the framework, as evident in Bache and Rasool (2006).

2.5.2 Representation of distributions

Adopting the scheme employed in Spicer and Pratsinis (1996a) sizes are represented in scaled form u = d/λ in which λ is an arbitrary length scale. Under this description the probability density function for the *number-size distribution* is defined by:

$$\Psi(u) = \frac{1}{N}\frac{dn}{du} \tag{2.145}$$

In this definition, N is the total number of flocs per unit volume and δn/N refers to the fraction found in the interval u to u + δu. In integrated form, this is consistent with the expression:

$$\int_{o}^{\infty} \Psi(u)\, du = 1 \tag{2.146}$$

In the particular case in which λ is represented by d_L (the arithmetic mean value of d across the number size distribution), u = d/d_L, this being consistent with the relationship:

$$\int_{o}^{\infty} u\Psi(u)\, du = 1 \tag{2.147}$$

Mass conservation is encapsulated by the identity:

$$M = \int_{o}^{M} dm \qquad (2.148)$$

For a suspension comprising quasi-spherical aggregates:

$$dm = \frac{\pi}{6} d^3 C(d)\, dn \qquad (2.149)$$

where C(d) is the average solids concentration *within* the floc. Making use of Equation (2.22) shows:

$$\frac{dm}{dn} = \frac{\pi}{6} d_o^3 A' \rho_s \left(\frac{d}{d_o}\right)^D \qquad (2.150)$$

The next task is to identify Ψ(u). Following Rosen (1984), a convenient choice is the gamma distribution:

$$\Psi(u) = Bu^a \exp(-bu^c) \qquad (2.151)$$

in which B, a, b and c are constants. When a > 0, this has the property that Ψ(u) → 0 when u → 0 and u → ∞ and possesses a single maximum (*i.e.* unimodal) in the intervening range. One advantage of Equation (2.151) is that it contains sufficient parameters to permit Ψ(u) to be matched (insofar as possible) to an observed distribution. Examples of the form of Ψ(u) are shown in Figure 3.22. A second advantage of Equation (2.151) is that it is readily integrated, noting that the gamma function Γ(α) is defined by:

$$\Gamma(\alpha) = \int_{0}^{\infty} t^{\alpha-1} e^{-t}\, dt, \qquad \alpha \geq 1 \qquad (2.152)$$

Substituting Equation (2.151) into Equations (2.146) and (2.147) leads to the definitions:

$$B = \frac{cb^{(a+1)/c}}{\Gamma((a+1)/c)}, \quad b = \left(\frac{\Gamma((a+2)/c)}{\Gamma((a+1)/c)}\right)^c \qquad (2.153), (2.154)$$

From this base, integration of Equation (2.150) leads to:

$$M = \frac{\pi}{6}\rho_s d_o^3 NA'S(D)\left(\frac{d_L}{d_0}\right)^D \tag{2.155}$$

where:

$$S(D) = \frac{\Gamma((a+D+1)/c)}{b^{D/c}\Gamma((a+1)/c)} \tag{2.156}$$

S(D) (a distribution moment) conveys information about the influence of the shape of $\psi(u)$ on the distribution of mass. Equation (2.155) shows how the solids mass concentration (M) can be expressed in terms of representative base variables. For example, if M is fixed (*i.e.* mass conserved) and G is altered, the product $NA'S(D)\ (d_L/d_o)^D$ must remain at a fixed value. Thus one can see interdependency between the base variables, including shape. The most obvious interdependence is the linkage between N and d_L (*e.g.* Figure 3.23). Examples of the way in which Equation (2.155) can be used to analyse data are described in Bache and Rasool (2006).

Table 3.12 summarises relevant information for representing distributions found in a flocculator at full scale; it is based on the framework described above.

2.5.3 Distribution moments

'Moments' are a useful means for expressing representative features of a distribution tied to its shape, the 'centre of gravity' or 'radius of gyration' being common examples. Formally, the pth moment is defined by:

$$< u^p > = \int_0^\infty u^p\, \Psi(u)\, du \tag{2.157}$$

Inspection of Equations (2.146) and (2.147) shows that these are distribution moments corresponding to the cases p = 0 and p =1 respectively; the former describes moments of the number-size distribution whereas the latter refers to the length-size distribution. The case p = 3 is of particular interest, because it focuses on the volume size distribution. The term $<u^3>$ represents the expression $(VAD/d_L)^3$ in which VAD is the *volume average diameter* (*i.e.* the arithmetic average of all cluster volumes across the distribution. When $\Psi(u)$ is represented by Equation (2.151), the pth moment is defined by:

$$< u^p > = \frac{1}{b^{p/c}} \frac{\Gamma((a+p+1)/c)}{\Gamma((a+1)/c)} \tag{2.158}$$

showing that the VAD/d_L ratio is specified by:

$$\frac{VAD}{d_L} = \frac{1}{b^c} \left(\frac{\Gamma((a+4)/c)}{\Gamma((a+1)/c)} \right)^{1/3} \tag{2.159}$$

Where percentiles are required *e.g.* the median value of the volume size distribution (d_{V50}) or its 95th percentile (d_{V95}), this requires evaluation of the incomplete gamma function, which is defined by:

$$\gamma(\alpha, x) = \int_0^x t^{\alpha-1} e^{-t}\, dt = x^\alpha e^{-x} \left[\frac{1}{\alpha} + \frac{x}{\alpha(\alpha+1)} + \frac{x^2}{\alpha(\alpha+1)(\alpha+2)} + \dots \right] \tag{2.160}$$

A partition of the moment distribution $u^p\Psi(u)$ depends on solution of the equation:

$$I(\alpha, x) = \gamma(\alpha, x) / \Gamma(\alpha) \tag{2.161}$$

in which I(α, x) signifies the desired partition fraction lying in the range (0, 1). When this is applied to the pth moment of Ψ(u), α and x are defined by:

$$\alpha \equiv (a + p + 1/c) \quad , \quad x \equiv bu^c \qquad (2.162), (2.163)$$

As an example, consider the calculation procedure for evaluating d_{V95} for a distribution represented by the parameters a = 11, b = 49.5, c = 0.25 (corresponding to data shown in Table 3.11). For the volume distribution p = 3, Equation (2.162) showing that α = 60. With I(α, x) = 0.95, the task is to find the value of x satisfying Equation (2.161) in the form γ(60,x) = 0.95Γ(60). By guessing values of x for substitution in the series approximation of Equation (2.160), the target solution *i.e.* 0.95Γ(60) can be found by a process of trial and error. For the example cited, the solution exists at x = 73.28. With this information Equation (2.163) yields u = 4.80 ($\equiv d_{V95}/d_L$) (see data at foot of Table 3.11).

Data summarised in Table 3.11 show a range of length scales based on moments of the source number-size distributions.

2.6 REFERENCES

Adler, P. M. and Mills, P. M. (1979) Motion and rupture of porous sphere in a linear flow field. *J. Rheol.*, 23(1), 25-37.

Bache, D. H. (2004) Floc rupture and turbulence: a framework for analysis. *Chem. Eng. Sci.*, 59, 2521-2534.

Bache, D. H., Johnson, C., Papavasilopoulos, E., Rasool, E. and McGilligan, F.J. (1999a) Sweep coagulation: structures, mechanisms, practice. *J. Water SRT-Aqua*, 48(5), 201-210.

Bache, D. H. and Rasool, E. (1996) Measurement of the rate of energy dissipation around an oscillating grid by an energy balance approach. *Chem. Eng. J.*, 66, 105-115.

Bache, D.H. and Rasool, E. (2006) Characterisation, scaling and analysis of steady state floc size distribution using a mass balance approach. *J. Water SRT-Aqua*, 55(5), 335-356.

Bache, D. H., Rasool, E., Ali, A. and McGilligan, J. F. (1995) Floc character: measurement and role in optimum dosing. *J. Water SRT – Aqua*, 44(2), 83-92.

Bache, D.H., Rasool, E., Moffat, D. and McGilligan, F. J. (1999b) On the strength and character of alumino-humic flocs. *Water Sci. Tech.*, 40(9), 81-88.

Bagster, D. F. and Tomi, D. (1974) The stresses within a sphere in simple flow fields. *Chem. Eng. Sci.*, 29, 1773-1783.

Balankin, A.S. (1996) Elastic behaviour of materials with multifractal structures. *Phys. Rev. B*, 53(9), 5438-5443.

Baldyga, J. and Bourne, J. R. (1995) Interpretations of turbulent mixing using fractals and multifractals. *Chem. Eng. Sci.*, 50(3), 381-400.

Batchelor. G.K. (1947) Kolmogoroff's theory of locally isotropic turbulence. *Proc. Cam. Phil. Soc.*, 43, 533-559.

Batchelor, G.K. (1951) Pressure fluctuations in isotropic turbulence. *Proc. Cam. Phil. Soc.*, 47, 359-374.

Bisplinghoff, R. L., Mar, J. W. and Pian, T. H. H. (1965) *Statics of deformable solids.* Addison-Wesley, New York.

Blaser, S. (2000a) Break-up of flocs in contraction and swirling flows. *Colloids and Surfaces A: Physicochem. Eng. Aspects,* 166, 215-223.

Blaser, S. (2000b) Flocs in shear and strain flows. *J. Colloid Interface Sci.*, 225, 273-284.

Boller, M. and Blaser, S. (1998) Particles under stress. *Water Sci. Tech.*, 39(10), 9-29.

Brinkman, H.C. (1947a) A calculation of the viscous force exerted by a flowing fluid on a dense swarm of particles. *Appl. Sci. Res. A*, 1, 27-34.

Brinkman, H.C. (1947b) A calculation of the viscosity and sedimentation velocity for solutions of large chain molecules taking account the hampered flow of the solvent through each chain molecule *Proc. R. Dutch Acad. Sci.*, 50, 618-625 and 821.

Buscall, R, Mills, P.D.A., Goodwin, J.W. and Lawson, D.W. (1988) Scaling behaviour of the rheology of aggregate networks formed from colloidal particles. *J. Chem. Soc., Faraday Trans. I*, 84(12), 4249-4260.

Buscall, R., Mills, P.D.A. and Yates, G.E. (1986) Viscoelastic properties of strongly flocculated polystyrene latex dispersions. *Colloids Surf.*, 18, 341-358.

Buscall, R. and White, L. R. (1987) The consolidation of concentrated suspensions. Part 1. – The theory of sedimentation. *J. Chem. Soc., Faraday Trans. I*, 83, 873-891.

Bushell, G. C., Yan, Y. D., Woodfield, D., Raper, J. and Amal, R. (2002) On techniques for the measurement of the mass fractal dimension of aggregates. *Adv. Colloid Interface Sci.*, 95, 1-50.

Chabra, A. and Jensen, R.V. (1989) Direct determination of the f(α) singularity spectrum. *Phys. Rev. Letters* 62(12), 1327-1330.

Channell, G.M. and Zukoski, C. F. (1997) Shear and compressive rheology of aggregated alumina suspensions. *AIChE J.*, 43(7), 1700-1708.

Clift, R., Grace, J.R. and Weber, M.E. (1978) *Bubbles, Drops and Particles*. Academic Press. New York.

des Cloizeaux, J. and Jannink, G. (1987) *Les polymers en solution: Leur modélisation et leur structure*. Les Editions de Physique, Les Ulis, France.

Drost-Hansen, W. (1977) Effects of vicinal water on colloidal stability and sedimentation processes. *J. Colloid Interface Sci.*, 58(2), 251-262.

Einstein, A. (1906) On the theory of Brownian movement. *Ann. Phys.* (Leipz), 19, 371-391.

Feder, J. (1988) *Fractals* Plenum Press, New York.

Firth, B. A. (1976) Flow properties of coagulated colloidal suspensions II. Experimental properties of the flow curve parameters. *J. Colloid Interface Sci.*, 57, 257-265.

Firth, B.A. and Hunter, R.J. (1976a) Flow properties of coagulated colloidal suspensions I. Energy dissipation in the flow units. *J. Colloid Interface Sci.*, 57(2), 248-256.

Firth, B.A. and Hunter, R. J. (1976b) Flow properties of coagulated colloidal suspensions III. The elastic floc model. *J. Colloid Interface Sci.*, 57(2), 266-275.

Fuchs, N. A. (1964) *Mechanics of Aerosols*. Pergamon, Oxford.

Gmachowski, L. (1996) Hydrodynamics of aggregated media. *J. Colloid Interface Sci.*, 178, 80-86.

Gmachowski, L. (1998) Flow drag in polydisperse systems. *Water Res.*, 32(9), 2655-2659.

Gmachowski, L. (2000) Estimation of the dynamic size of fratal aggregates. *Colloids Surf. A: Physicochem. Eng. Aspects*, 170, 209-216.

Gmachowski, L. (2001) A method of maximum entropy modelling the aggregation kinetics. *Colloids Surf. A: Physicochem. Eng. Aspects*, 176, 151-159.

Gmachowski, L. (2002a) *Hydrodynamic behaviour of suspensions. Encyclopedia of Surface and Colloid Science*. Marcel Dekker, New York, 2364-2374.

Gmachowski, L. (2002b) Comment on "Permeability of fractal aggregates" *Water Res.*, 36, 3416-3417.

Gmachowski, L. (2003) Mass-radius relation for fractal aggregates of polydisperse particles. *Colloids Surf. A: Physicochem. Eng. Aspects*, 224, 45-52.

Grout, H., Tarquis, A.N. and Wiesner, M.R. (1998) Multifractal analysis of particles size distributions in soil. *Environ. Sci. Technol.*, 32, 1176-1182.

Happel, J. (1958) Viscous flow in multiparticle systems: slow motion of fluids relative to beds of spherical particles. *AIChE. Journal* 4, 197-201.

Happel, J. and Brenner, H (1965) *Low Reynolds Number Hydrodynamics: with Special Applications to Particulate Media*., Prentice Hall, Inc.,Englewood Cliffs, N.J., USA.

Herweijer, J. A. (1995) *The Small-scale Structure of Turbulence*. Technische Universiteit Eindhoven, Eindhoven. ISBN 90-396-0137-9.

Hill, R. and Power, G. (1956) Extremum principles for slow viscous flow and the approximate calculation of drag. *Quart. J. Mech. Appl. Math.*, 9, 313-319.

Hinze, J.O. (1955) Fundamentals of the hydrodynamic mechanism of splitting in dispersion processes. *AIChE Journal,* 1(3), 289-293.

Jeffrey, D. J. (1982) Aggregation and break-up of clay flocs in turbulent flow. *Advances in Colloid and Interface Sci.*, 17, 213-218.

Kolmogorov, A.N. (1941) The local structure of turbulence in incompressible viscous flow for very large Reynolds numbers. *C.R. Acad. Sc. U.R.S.S.*, 30, 309.

Ladyzhinsky, I. Ya. and Uriev, N. B. (1996) Whether or not two types of layer coexist in colloidal gels under shear. *Colloids Surfaces A: Physiochem. Eng. Aspects*, 108, 13-25.

Lagvanker, A. L. and Gemmell, R. S. (1968) A size-density relationship for flocs. *J. Am. Water Works Assoc.*, 60(9), 1040-1046.

Lawler, D. F. (1997) Particle size distributions in treatment processes: theory and practice. *Water Sci. Tech.*, 36(4), 15-23.

Mandelbrot, B.B. (1977) *Fractals, Form, Chance and Dimension.* W.H. Freeman, San Francisco.

Marin, G. (1988) Oscillatory rheometry. In: *Rheological Measurement* (Eds. Collyer, A. A. and Clegg, D.W.) Elsevier Applied Science, London, 297-343.

Michaels, A.S. and Bolger, J.C. (1962) The plastic flow behaviour of flocculated kaolin suspensions. *Ind. Eng. Chem. Fundam.*, 1, 153-162.

Moon, F.C. (1992) *Chaotic and Fractal Dynamics* John Wiley & Sons, Inc., New York.

Neale, G., Epstein, N. and Nader, W. (1973) Creeping flow relative to permeable spheres. *Chem. Eng. Sci.*, 28, 1865-1874.

Papenhuijzen, J. M. P. (1972) The role of particle interactions in the rheology of dispersed systems. *Rheol. Acta*, 11, 73-78.

Paunov, V. N., Dimova, R. I., Kralchevsky, P.A., Broze, G. and Mehreteab, A. (1996) The hydration repulsion between charged surfaces as an interplay of volume exclusion and dielectric saturation effects. *J. Colloid Interface Sci.*, 182, 239-248.

Piau, J-M., Dorget, M. and Palierne, J-F. (1999) Shear elasticity and yield stress of silica-silicone physical gels: Fractal approach. *J. Rheol.*, 43(2), 305-314.

Prestidge, C. and Tadros, T. F. (1988) Rheological investigation of depletion flocculation of concentrated sterically stabilised polystyrene latex dispersion. *Colloids Surf.*, 31, 325-346.

Probstein, R.F. (1994) *Physicochemical Hydrodynamics: An Introduction.* Second Edition. John Wiley & Sons, New York.

Rees, A. L. G. (1951) Directed aggregation in colloidal systems and the formation of protein fibres. *J. Phys. Colloid Chem.*, 55, 1340-1344.

Robinson, J. and Knocke, W.R. (1992) Use of dilatometric and drying techniques for assessing sludge dewatering characteristics. *Water Environ. Res.*, 64(1), 60-68.

Rosen, J.M. (1984) A statistical description of coagulation. *J. Colloid Interface Sci.*, 99(1), 9-19.

Sato, H., Eto, S. and Susuki, H. (1982) Dewatering sludges: relationship between amount of bound water and dewatering characteristics of alum sludges. *Filtr. Separat.*, 19(6), (Nov/Dec), 492-497.

Shaw. D.J. (1992) *Introduction to Colloid and Surface Chemistry.* Fourth Edition. Butterworth-Heinemann, Oxford.

Shinnar, R. (1961) On the behaviour of liquid dispersions in mixing vessels. *J. Fluid Mech.*, 10, 259-275.

Sonntag, R.C. and Russel, W.B. (1987) Structure and breakup of flocs subjected to fluid stresses. II. Theory. *J. Colloid Interface Sci.*, 115(2), 378-389.

Sonntag, H and Strenge, K (1987) *Coagulation Kinetics and Structure Formation.* Plenum Press, New York.

Sorensen, C.M. and Roberts, G.C. (1997) The prefactor of fractal aggregates. *J. Colloid Interface Sci.*, 186, 447-452.

Spicer, P.T. and Pratsinis, S.E. (1996a) Shear-induced flocculation: the evolution of floc structure and the shape of the size distribution at steady state. *Water Res.*, 30(5), 1049-1056.

Spicer, P.T. and Pratsinis, S.E. (1996b) Coagulation and fragmentation: universal steady-state particle size distribution. *AIChE Journal*, 42(6), 1612-1620.

Stauffer, D, (1979) Scaling theory of percolation clusters. *Phys. Rep. (Review Sect. Phys. Lett.)* 54(1), 1-79.

Tambo, N. and Hozumi, H. (1979) Physical characteristics of flocs – II. Strength of floc. *Water Res.*, 13, 421-427.

Tambo, N. and Watanabe, Y. (1979) Physical characteristics of flocs-I. The floc density function and aluminium floc. *Water Res.*, 17, 409-419.

Tanaka, H. Nishikawa, Y. and Koyama, T. (2005) Network-forming phase separation of colloidal suspensions. *J. Phys.: Condens. Matter*, L143-L153.

Tang, P. and Raper, J. A. (2002) Modelling of the settling behaviour of fractal aggregates – a review. *Powder Tech.*, 123, 114-125.

Torres, F. E., Russel, W. B. and Schowalter (1991) Floc structure and growth – kinetics for rapid shear coagulation of polystyrene. *J. Colloid Interface Sci.*, 142(2), 554-574.

Uriev, N.B. and Ladyzhinsky, I. Ya. (1996) Fractal models in rheology of colloidal gels. *Colloids Surfaces A: Physiochem. Eng. Aspects*, 108, 1-11.

van de Ven, T. G. M. and Hunter, R.J. (1977) The energy dissipation in sheared coagulated sols. *Rheol. Acta*, 16, 534-537.

van den Tempel, M. (1961) Mechanical properties of plastic-disperse systems at very small deformations. *J. Colloid Sci.*, 16, 284-286.

Veerapaneni, S. and Wiesner, M.R. (1996) Hydrodynamics of fractal aggregates with radially varying permeability. *J. Colloid Interface Sci.*, 177, 45-57.

Von Smoluchowski, M. (1916) Drei vortrage uber diffusion, Brownsche Molekularbewegung und koagulation vor kolloidteilchen. *Phys. Z.*, 17, 557-571.

Von Smoluchowski, M. (1917) Versuch einer mathematischen Theorie der Koagulationskinetic kolloider Lösungen. *Z. Phys. Chem.*, 92, 129-168.

Woodfield, D. and Bickert, G. (2001) An improved permeability model for fractal aggregates settling in creeping flow. *Water Res.*, 35(16), 3801-3806.

Yan, Yao-de, Burns, J. L., Jameson, G. J. and Biggs, S. (2000) The structure and strength of depletion force induced particle aggregates. *Chem. Eng. J.*, 80, 23-30.

Yeung, A. K. C. and Pelton, R. (1996) Micromechanics: A new approach for studying the strength and breakup of flocs. *J. Colloid Interface Sci.*, 184(2), 597-585.

Section 3 Flocs in Practice

3.1 MEASUREMENT

3.1.1 Perspective

In general, the task of measuring floc properties is beset with difficulties stemming from the combined influences of composition, form, structure and rheology. Some problems can be reduced by using model suspensions, but much of the complexity remains when using hydrolysing metal salts that originate from the principal coagulant. For example, how does the presence of a particulate gel within the floc interstices influence settlement or the optical properties?

Philosophically there are questions that are always faced in measurement *e.g.* how does one know that the measurement is 'right'? This is very pertinent in size measurement. "…but I calibrated the system with standard particles…" (a conference quote). Flocs are not standard particles! Instruments have a great capacity to yield information, but what does it mean? There are no ready answers to these questions. Where possible, measurements should be cross-checked by independent means. More often than not, one gets different versions of 'the truth'. Interpretation demands a good understanding of the materials which are being examined and how a measurement system responds to the particular system under investigation.

Here, we address some of the common problems encountered when measuring size, size distributions and fractal properties. Readers are referred to Elimelech *et al.* (1995, Chapter 9) for a wider and more detailed view of techniques for measuring aggregates.

3.1.2 Size

3.1.2.1 Some experiences

There was an occasion when the author (DHB) took an 'off-the-shelf' particle size detector (PSD) to a waterworks. Flocs were sampled and subjected to shear in a stirred beaker. Visually all seemed well until the flocs (alumino-humic from a low turbidity water) were fed into a detection cell within the instrument. The PSD generated information about size on the basis 'interception times' when particles were sensed by a rotating laser beam *e.g.* Williams *et al.* (1992).

ISBN: 9781789065442. Published by IWA Publishing, London, UK.

The output showed little, bar the occasional particle. The problem was caused by the optical transparency of the flocs. This type of instrument responds better when the base suspension is formed from larger 'solid' particles (*i.e.* with higher refractive index). When faced with this type of problem, *a useful test* is to see how the instrument responds to a freshly prepared aluminium hydroxide suspension. Under controlled mixing, flocs (barely visible) should exhibit behaviour which is roughly in accord with the findings reported in Bache *et al.* (1999b).

A second problem described in Bache *et al.* (2000), centres on the arrangement shown in Figure 3.1. In this system, flocs were measured remotely via a low power microscope coupled to a CCTV system - the recordings being subjected to image analysis. Measurements were made in an illuminated plane adjacent to the edge of a two-litre Phipps and Bird square beaker, and across a cuvette (10x10x40 mm) which was designed to minimise disturbances in a sample stream from the beaker.

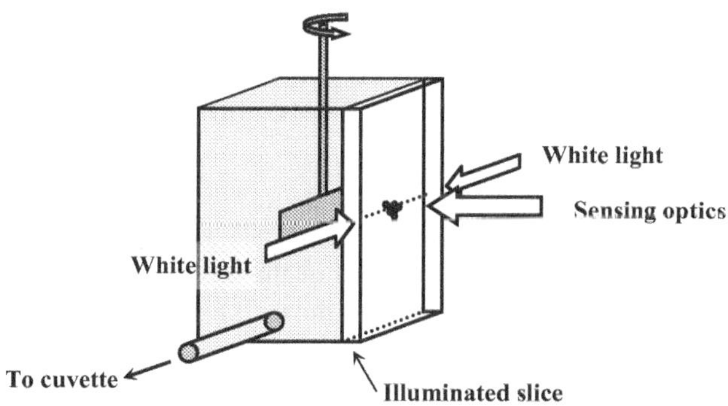

Figure 3.1. Mixing, sampling and measurement system associated with Figure 3.2.

Figure 3.2 shows a comparison of the floc sizes detected at the two sampling locations. It is evident that major discrepancies exist between the two systems of measurement, particularly at low G values. Two explanations were proposed in Bache *et al.,* 2000). The first of these was that flocs were being ruptured as a result of shear forces during transport to the cuvette. A second and quite different explanation was that during stirring, the equilibrium size distribution (arising from continuous formation and rupture) was larger than when these kinetic processes were absent. Subsequent evidence (Bache and Rasool, 2006) appears to lend credence to this explanation. However, it seems more likely that the larger and weaker flocs were being ruptured within the sample stream. Many lessons can be drawn from this example. First, the necessity of cross-checking measurement techniques. Second, when a sample stream containing flocs is fed into a PSD system, it is vital to appraise the potential impact of the internal stress regime on the materials being measured. Third, when studying 'large flocs', remote sensing techniques are advocated.

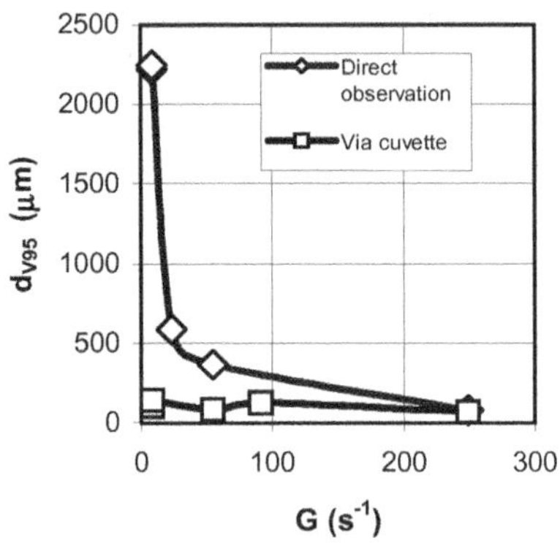

Figure 3.2. Comparison of maximum floc size – velocity gradient variation in sample stream and at remote position illustrated in Figure 3.1 (Bache et al., 2000).

3.1.2.2 Stress regimes in PSD

When a sample stream passes through a PSD system, flocs may experience stress conditions that are well in excess of the environment in which they were created. Critical points include passage through a sensing orifice or flow cell or in contraction flows. Useful information can be gleaned from knowledge of the pressure losses (ΔP) across critical stages, these being translated into an average velocity gradient using the formula:

$$G = \sqrt{\frac{Q\,\Delta P}{\mu V}} \tag{3.1}$$

in which Q is the volumetric flow rate and V the volume of the measurement section. It should be noticed that V/Q is the average retention time. For example in a Coulter counter (reviewed in Elimelech *et al.*, 1995) with a 100 μm orifice, manufacturer's data indicates ΔP = 20333 Pa (corresponding to 6 inches of Hg), orifice length: width = 1:0.75 and Q ~ 30 μ*l* s^{-1}. If the sensing volume is just based on the dimensions of the orifice (probably an underestimate), this translates into G > 50,000 s^{-1}. This has the potential to severely alter particle size counts in the larger size ranges as reported in Gibbs (1982a).

In the case of a flow cell in a HIAC 9703 (a PSD system), manufacturer's data (Hiac-Royco, USA) indicates that the cell has a cross-sectional area of 0.4 mm^2 and transports a sample flow of 1 ml s^{-1}. This corresponds to a speed of 2.5 m s^{-1} through the cell. This suggests that flocs within the sample stream will encounter zones of large acceleration and shear – a recipe for rupture *e.g.* Gibbs (1982b). This is borne out by data shown in Figure 3.3. This compares the distribution measured by the HIAC with one obtained by remote sensing. The measuring systems provide different perspectives of the size distribution.

In contrast to the HIAC 9703, the remote system shows the presence of a significant fraction of the floc numbers above 30 μm, suggesting that these were ruptured in the HIAC system. The image analysis system was unable to provide information about the distribution below 9 μm (the lowest limit of resolution). Further, when tested against standard particles, size measurements below 20 μm were unsatisfactory (a resolution problem for the particular system in use).

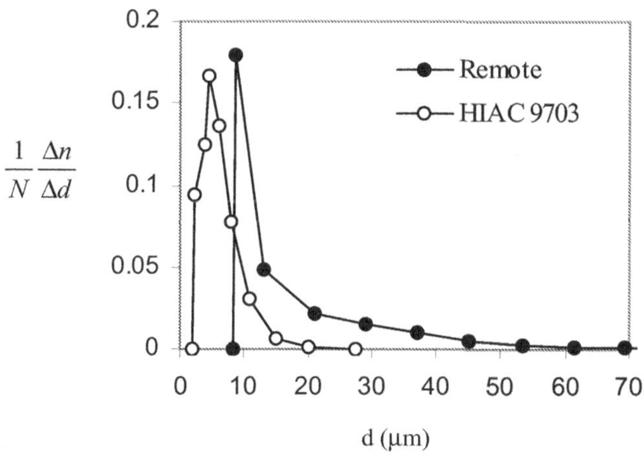

Figure 3.3. Comparison of floc size distributions in a stirred humic suspension coagulated with aluminium sulphate using remote sensing and via a flow cell in a HIAC 9703. The lower cut-off values correspond to the respective instrument resolution limits (Unpublished data, DHB).

3.1.2.3 Remote sensing

Remote sensing is strongly recommended for situations in which invasive sampling leads to significant perturbation of the system under investigation. Necessary components include the illumination system, microscope, digital camera and image analysis software. Amongst these, illumination is a major consideration. The configuration adopted depends on the vessel holding the flocs, the optical properties of the floc, and on the measurement objectives. Tasks may include detailed shape analysis of individual flocs or the measurement of size distributions and concentration *i.e.* the number of flocs per unit volume. Of these, the last is the most difficult because of the mix of factors involved (see Bache and Rasool, 2006).

Foreshortening a lengthy discussion, the author (DHB) has spent much time exploring in-line illumination, transverse illumination, and different forms of light source including lasers. When examining the effect of shear on optically transparent flocs, such as reported in Bache *et al.* (2000), the most satisfactory option was the arrangement illustrated in Figure 3.1 *i.e.* transverse illumination using a white light/broad band source. In these circumstances, laser illumination generates a 'speckled' appearance with strong variations in light intensity across the floc. In the case of an underwater camera used at full-scale, Bache *et al.* (2000) described an arrangement in which the focal plane was illuminated transversely using a ring of LEDs. When flocs are opaque, the measurement task is more straightforward, and there is greater freedom of choice of the illumination system.

By way of comment on the core theme in this Section, invasive sampling is acceptable in situations in which measurements are not unduly distorted by the measurement system. Also, where consistent errors are introduced (whatever the system of measurement) the resulting information can be of great value where one is seeking information about a relative response to a control variable, rather than information that is accurate in absolute terms.

3.1.2.4 Image analysis

When using the illumination arrangement illustrated in Figure 3.1, one can discriminate between objects 'in focus' and 'out of focus' on the basis of their relative light intensity; this is represented by a standard grey scale with 0 representing black (no light) and 255, the maximum brightness. Even so, there are many imponderables. Figure 3.4 shows the response of a floc image to different settings of the threshold setting (a selected level of brightness on the grey scale; see caption for implications). At the lowest setting (200) there is a single object, whereas at 255 (showing the brightest parts), there are six objects, even though the image frame refers to the same floc! Between these extremes the object number and size depends on the selected threshold setting. The selection of the threshold setting is a matter of judgment. From this, it is evident that there is an element of ambiguity in this approach to size measurement. In this respect it is vital to maintain a consistent approach to illumination and threshold selection across a sequence of measurements.

Judgement also has to be exercised on matters of image overlap *e.g.* is it a large floc or two overlapping flocs? The lower limit of resolution is set by the pixel area, the magnification factor and the minimum number of pixels used to define an image. Much depends on the type of analysis to be performed on an image (see Chakraborti *et al.*, 2000 & 2003). If it is simply to determine the cross-sectional area (from which one can calculate an equivalent diameter), as few as 6 pixels may be adequate *vis à vis* other uncertainties. Generally, it is vital to test the adequacy of the measurement system by measuring mono-sized suspensions of reference spheres of differing size. Video recordings at standard speed cannot cope with 'fast-moving' images at high magnification. In this respect, the system described by Chakraborti *et al.* (2000) has merit *i.e.* using a computer controlled digital camera in association with stroboscopic lighting.

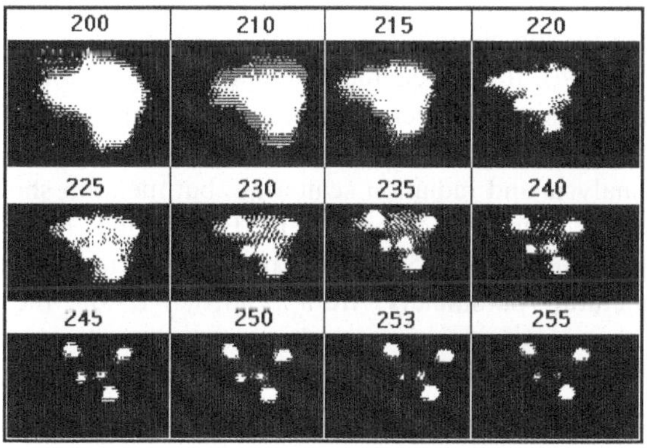

Figure 3.4. Illustrating response of floc image to light threshold level; white pixels in the image correspond to part of the illuminated floc that have light intensities equal or greater than the light intensity set by the threshold level. Below the threshold, pixels are made black (Bache *et al.*, 2000).

3.1.3 Fractal properties

3.1.3.1 Perspective

The review of Bushell *et al.* (2002) identifies the range of approaches for measuring fractal properties. Of the methods available, image analysis, sedimentation and light scattering appear to offer the best prospects for gaining information about floc structures. Of these, it must be recognised that image analysis is necessarily set in the context of two-dimensional space, whereas the others reside in three-dimensional space. In general, it is unclear how to translate information about fractal properties from one context to another.

For example, length-area relationships gained from the analysis of a two-dimensional image may display fractal behaviour, but this type of information cannot be transferred to the mass fractal dimension (on which sedimentation and scattering properties depend).

Our principal focus is on the fractal properties as embodied in either Equation (2.14) or Equation (2.26). In these, it is evident that information must be gained about *both* A′ and D, the latter referring to the *mass fractal dimension*. When this model is applied to regimes of behaviour within an object with a multifractal structure (see Figure 2.3), one must anticipate values of A′ which may be very different from unity as noted in Section 2.1.3.2. Indeed, the task of representing a multifractal structure is exceedingly complex. It should also be noted that the magnitude of A′ also depends on the selected values of d_o or δ_o. Sedimentation analysis has the capacity to generate information about both A′ and D, whereas radiation scattering analysis (as currently developed) can only provide information about the mass fractal dimension (or to be precise, the correlation dimension). It has become fashionable to gain information about D using scattering analysis. However, this is of limited value in terms of defining a fractal structure, unless one has knowledge of A′. The foregoing is confined to sedimentation analysis and radiation scattering, but mention should be made of the mass balance approach described in Bache and Rasool (2006) and encapsulated in Equation (2.155). This provides a means of checking or estimating structural parameters from knowledge of the number size distribution.

3.1.3.2 Sedimentation analysis

Sedimentation analysis is widely used for gaining information about fractal structure, but it is beset with difficulties. The basis of the method is to transform knowledge of the velocity-size dependence into a size-density dependence, *e.g.* Equation (2.24) from which estimates of A′ and D can be gained. The technique is time-consuming and demands a high degree of experimental rigour. Generally, it is an invasive technique, on the grounds that flocs must be removed from-, or experience conditions that are different from the dynamic environment in which they are formed. Background theory is discussed in Section 2.2.3. When interpreting data *e.g.* using Equation (2.67), one must resolve the problems posed by shape factors and permeability.

Temperature control

Floc density measurement must be carried out in a temperature controlled sedimentation column using the type of approach described in Bache *et al.* (1995). Temperature differences as small as 0.1 °C cause density differences that may induce unwanted convection. Spurious sources of heat include light sources (which must be used sparingly), and radiant energy from the observer. When transferring flocs into the column using a dipping tube, the sampled water containing the flocs must be less dense than the column water so that the bulk sample floats. Flocs are then able to settle through this region to deeper levels within the column. The consequences of introducing sample water with a density greater than that existing in the column are potentially disastrous in terms of generating meaningful information about floc density. The relative density of the sample water and the column water can be controlled by maintaining a temperature difference. For example, in the measurements reported in Bache *et al.* (1995), a temperature difference 2 °C was used. When dealing with very low density flocs with settling speeds ~ 10^{-4} m s^{-1}, even greater care must be exercised. In an investigation devoted to characterisation of flocs gained from humic waters (very low density differences), Bache *et al.* (1999b) found it advantageous to maintain the settling column at about 7 °C below the ambient temperature (15 °C). Prior to use, the column was left for at least 12 hours 'to stabilise'. Once floc samples had been transferred to the column, no further samples were added for about two hours to allow the column to 'restabilise'. In spite of these precautions, only 1/3 of the measurement sequences were successful, convection being evident in the remainder.

Shape factors

It is difficult to assign many of the shape factors reported in the literature, because one only has a two-dimensional view of the floc. The route advocated in Bache *et al.* (1995) is the Hill and Power (1956) approach noted in Section 2.2.3 and discussed in Bache and Hossain (1991). The maximum and minimum dimensions of a floc provide estimates of the diameter of the circumscribed and inscribed sphere respectively. Thereafter d is estimated using $d = \sqrt{(d_{min}d_{max})}$ which provides and estimate of the 'correct' answer. Alternatively one may measure the cross-sectional area A_c and obtain d from $d = \sqrt{(4A_c/\pi)}$. When one measures the settlement of a hundred or so flocs under a particular set of conditions, errors generated by these measures tend to balance out. Under this scheme, $f_s = 1$, because one has represented the floc by an equivalent sphere. With this simplification, the settlement in the domain Re < 1 complies with:

$$u = \frac{1}{\Omega} \cdot \frac{gd^2 \Delta\rho}{18\mu} \tag{3.2}$$

Permeability

The first hurdle in assigning a value to Ω is deciding whether the floc behaves as impermeable (Ω=1) or permeable (Ω<1). It is often assumed that flocs are impermeable, when in reality they may be permeable. This leads to errors in the evaluation of the density difference. The mutual dependence can be represented by the function $\Delta\rho(\Omega)$. When a floc is porous, the true density difference, $\Delta\rho(\Omega<1)$, is then defined by:

$$\Delta\rho(\Omega \leq 1) = \Omega\, \Delta\rho(\Omega = 1) \tag{3.3}$$

the impermeable equivalent being represented by $\Delta\rho(\Omega=1)$. Equation (3.3) shows that the assumption Ω = 1 overestimates the true density difference of a porous floc. When $\Delta\rho$ data is fitted by Equation (2.24), this has the potential to provide misleading information about both A′ and D. Numerical analysis reported in Wu and Lee (1998) suggests that D is insensitive to the porosity term, implying that errors in the size-density dependence are carried by A′. Employing the same notational scheme as above, we may write:

$$A'(\Omega \leq 1) = \Omega\, A'(\Omega = 1) \tag{3.4}$$

showing the impact of $\Omega(\leq 1)$ on A′.

In principle, if one knows the floc solids volume concentration and thereby the permeability *e.g.* via Equation (2.57), one can evaluate Ω using Equation (2.69). However the solids volume concentration (ϕ) is unique to a particular size or floc. The best estimate of ϕ, is usually provided by the density difference, which cannot be determined unless one knows Ω! Thus one ends up with a fairly intractable problem.

Porosity effects can be identified in various ways, for example by following the pathway of tracers through individual flocs. Some studies *e.g.* Lee *et al.*, (1996) have immobilized the fluid in the pore space by freezing the floc, and have identified porosity by comparing the settling speeds in the frozen and unfrozen state. Here, there must be concerns about the impact of temperature on the settlement process. Indications of porosity effects can be picked up by inspection of the size-density plots, an example being shown below.

Figure 3.5 refers to alumino-humic flocs strengthened by polymer addition. Data is well fitted by Equation (2.24) yielding the values A′ = 6.57×10^6, D = 1.14 when d_o = 8 nm (discussed in Bache *et al.*, 1999a) and ρ_s = 2000 kg m^{-3} (Hossain and Bache, 1991). When d < 50 μm, $\Delta\rho$ > 1000 kg cm^{-3}. For the materials involved, the maximum possible density difference will be about 1500 kg m^{-3} (based on coagulant products having a maximum density of about

2500 kg m^{-3} (Hossain and Bache, 1991). This estimate ignores the presence of bound water (which must exist) and will reduce the maximum effective density (Equation 2.31). Given these features, one suspects that the calculated values of $\Delta\rho$ (based on the assumption $\Omega = 1$) are 'high', pointing to the impact of permeability. Thus it appears that A′ is overestimated. In spite of this, its order of magnitude is correct.

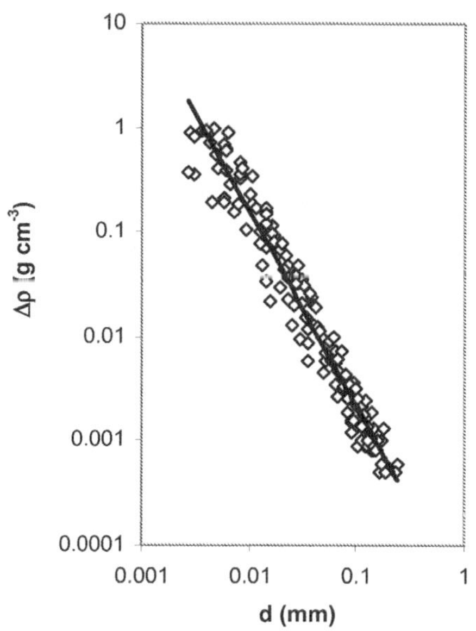

Figure 3.5. Example of size-density function for alumino-humic flocs strengthened by polymer addition.

A second example shows how order of magnitude calculations can provide insight into the factors involved. This refers to rice starch flocs reported in Figure 3 of Bache *et al.* (1995). In this case, a 250 mg/l suspension of rice starch was coagulated with alum at a dose of 10 mg Al/l. The mode of destabilisation was sweep floc, individual flocs envisaged as having the schematic composition illustrated in Figure 1.9. If water is to flow through the floc, then this must be through the pores of the space filling hydroxide gel. The size-density plot shows that the lowest density flocs were characterised by $\Delta\rho(\Omega{=}1) \approx 7$ kg m^{-3} with d = 500 μm. The density of rice starch (*i.e.* ρ_s, the dominant material) was 1500 kg m^{-3}, showing that $c = \Delta\rho/(1-\rho_w/\rho_s) \approx 21$ kg m^{-3} (Equation 2.23), c referring to the concentration of the solids mass within the floc. In flocs of this type, the size-density relationship was shown by Tambo and Watanabe (1979) to be characterised by the prevailing ALT ratio (Section 3.2.3). From information given above, ALT =10/250 = 0.04. Assuming that the hydroxide is in the form of $Al(OH)_3$, the equivalent mass of the precipitate is guided by the molecular ratio $[Al(OH)_3]/[Al]$ = 2.9. With this information, the concentration of precipitate within the floc is about 2.4 kg m^{-3} (= 21x 0.04 x 2.9). Noting that the precipitate has a density close to 2500 kg m^{-3} (Hossain and Bache, 1991), its solids volume concentration within the floc is $\phi \sim 10^{-3}$. Reference to Figure 2.6 shows $U \approx 0.9$ when $\phi = 0.001$. Taking hydroxide particles as having a diameter of about 8 nm, evaluation of Equation (2.57a) shows that the gel has a permeability ~3000 nm^2. Substituting this value into Equation (2.70) shows $\beta \sim 4000$ and $\Omega = 1$ (via Equation 2.69). The calculations suggest that the pore space behaves as if it is filled by an impermeable jelly. This supports the original assumption (Ω = 1) and is consistent with the findings of Matsumoto *et al.* (1978) for bentonite-alum flocs. When these circumstances apply, estimates of A' should be correct.

Before leaving this subject, it is worth comparing the two examples *i.e.* why the alumino-humic flocs associated with Figure 3.5 appear to be porous, whereas the rice starch flocs discussed above behave as be impermeable. In both cases, the flocs have a similar space-filling material *i.e.* hydroxide gel. One is forced to conclude that the polymer used for floc strengthening alters the spatial uniformity of the gel - thus creating larger pore spaces through which water can flow. This view is supported by the effects of polymer on dewatering of hydroxide sludges (Section 3.3.7; Table 3.29).

3.1.3.3 Scattering

When a well-collimated beam of monochromatic radiation (light, X-rays or neutrons) is incident on a sample, some of the radiation will pass though the sample unaffected (*i.e.* transmitted) or at the other extreme may be absorbed. Between these extremes, a certain fraction will interact with the sample, emerging as *scattered radiation* at an angle $\theta > 0$. By analysing the manner in which radiation is scattered, information can be gained about the size, form and structure of the materials causing the scatter. Scattering is attributed to 'fluctuations' in the medium, these being associated with variations in the density of the scattering material within the medium. Colloidal particles scatter light when their refractive index differs from the liquid in which they are dispersed.

The basic procedure in scattering measurement is to determine the intensity of the scattered radiation, $I(\theta,t)$, using a detector at a distance away from the sample as illustrated in Figure 3.6. In a dilute solution, the scattering pattern provides information about individual particles and the arrangement of material within them, whereas in a more concentrated solution, one can gain information about the average spatial arrangement of particles in a sample.

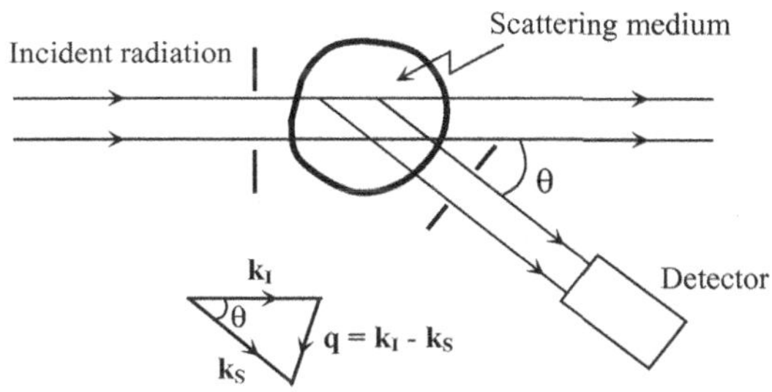

Figure 3.6. Schematic layout for scattering measurement, showing also the vectors for propagation and scattering (adapted from Pusey, 2004).

The principles of scattering are similar for the different forms of radiation, but there are distinct underlying differences. Light scattering is associated with the dielectric properties (or refractive index), whereas X-rays are scattered by electrons, or in the case of neutrons, by the nuclei of atoms. These differ in terms of the energy delivered (or penetrating power) and wavelength.

Monochromatic radiation propagating in the x direction can be represented by the waveform $E = E_o \cos(\omega t - kx)$ in which k is the wave number, ω the frequency and E_o the amplitude of the electric field. Parameter x depicts the distance of travel. In the general case, information about direction is expressed in the form of a propagation vector (**k**). For example, the incident radiation on a sample can be represented by the propagation vector ($\mathbf{k_I}$) of magnitude $|\mathbf{k_I}| = k_I = 2\pi/\lambda$ where λ is the wavelength of the light in the medium (as distinct from its value in a vacuum).

The wave number scales the number of 'wave lengths' per unit distance. When light is scattered, its new path is represented by the propagation vector $\mathbf{k_S}$. Reference to Figure 3.6 shows that the scattering vector **q** is defined as the difference between the propagation vectors for the incident and scattered light. A major simplification can be achieved for situations in which the energy losses arising from the interaction of radiation with matter negligible.

This is referred to as *quasi-elastic scattering* and means that we can ignore changes in both wavelength and frequency that may result from scattering. In these circumstances the magnitude k_S of the vector $\mathbf{k_S}$ is also $2\pi/\lambda$. When this occurs, the geometry of the vector triangle (Figure 3.6) shows:

$$q = |\mathbf{q}| = \frac{4\pi}{\lambda}\sin\frac{\theta}{2} \tag{3.5}$$

Equation (3.5) indicates that the magnitude of **q** is a function of both the scattering angle and the wavelength of the radiation. It should be recognised that q is a quantity set in 'wave number space'. The association between q and distances (l) set in 'real space' can be roughly estimated using the rule of thumb $l \cong 2\pi/q$ (Pusey, 2004). It is useful to envisage l as the diameter of a probing sphere within which information is gained about the scattering behaviour. Its diameter is controlled by q, and thereby by the scattering angle. Table 3.1 shows approximate values of the scattering vector for various scattering angles and different forms of radiation. Because of their short wave length and greater energy, X-rays and neutrons can probe structures that are much smaller than those accessible to light. Information about the largest scales of interest depends on the behaviour of scatter patterns when θ is small.

Table 3.1. Approximate magnitudes of the scattering vector and its reciprocal for X-rays, neutrons and light at various scattering angles (from Pusey, 2004).

X-rays, neutrons (λ = 0.1 nm)			*Light (λ = 400 nm)*		
θ (degrees)	q (nm^{-1})	$2\pi/q$ (nm)	θ (degrees)	q (nm^{-1})	$2\pi/q$ (nm)
0.01	0.01	600			
0.1	0.1	60			
1	1	6	1	3×10^{-4}	2×10^{4} (20 μm)
10	10	0.6	10	3×10^{-3}	2000
			100	2.4×10^{-2}	300

The conditions for quasi-elastic scattering are realised when scattering is very weak. 'Weakness' implies that the electric field of the scattered radiation is largely unaffected by the presence of the scattering particles. In accord with the Raleigh-Gans-Debye (RGD) approximation (*e.g.* Glatter, 2004), this occurs when $d \ll \lambda$, *i.e.* when the scattering particles are much smaller than the incident wavelength. When this condition applies, the scattering centres act independently. To obtain the overall intensity of scattered light at a detection point, we simply add up the waves arriving from the individual scattering centres, taking account of phase shifts arising from different path lengths through the sample.

Figure 3.7 shows the association between scattering methodology and particle size. When using light, the largest particle dimension satisfied by the RGD approximation is about 100 nm. The RGD approach is also applicable to neutrons and X-Ray scattering. When analysis is based on analysis of the angle of the average scattered intensity, this is referred to as *static scattering*. If particles are moving, one may also carry out *dynamic scattering* - the important element being the inclusion of time. When particle sizes are large (say $d > 40\lambda$) Fraunhofer diffraction theory is applicable.

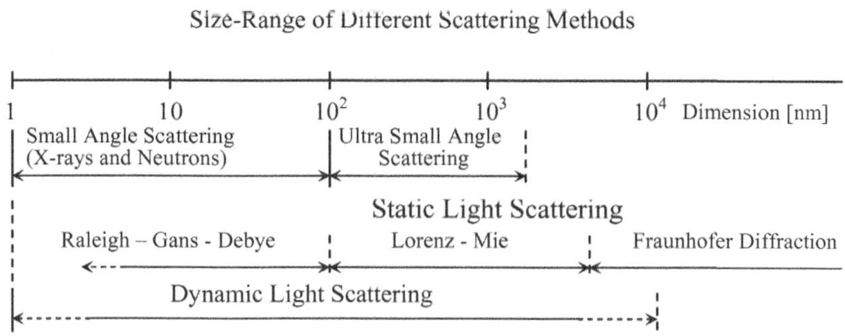

Figure 3.7. Range of applicability of the scattering methods (adapted from Glatter, 2004).

For example, it applies to particle diameters > 16 μm when light wavelength of 400 nm is used. Clearly, this is above the colloidal range. The intervening 'Lorenz-Mie' range addresses the problem of scatter when $d \sim \lambda$. Generally it takes account of the complex interaction between light and matter. Its great asset is that it permits the estimation of particle size distributions in the range 50 nm to a few microns on the basis of the scattering behaviour. At the present time, it does not appear to lend itself to structure analysis.

Scattering theory (RGD)

Scattering theory is expressed in many different ways. The following is based on Pusey (2004). For a dilute suspension with N particles per unit volume (V), the average intensity of the scattered radiation is expressed by:

$$< I(q) > = (E_o^2 / R^2)\, N|b(\mathbf{q})|^2 \tag{3.6}$$

Parameter R accounts for the distance from a reference origin and b(**q**) refers to the *scattering length* of an individual particle. The term <...> represents an average over time, with **q** defined in Figure 3.6. This can be written in the form:

$$< I(q) > \propto N|b(0)|^2 P(q) \quad \text{with} \quad P(q) = \frac{|b(\mathbf{q})|^2}{|b(0)|^2} \tag{3.7}$$

where P(q) is called the *form factor* of the particle. It is defined so that $P(q) \rightarrow 1$ as $q \rightarrow 0$ and provides information about the structure of individual particles. The term $|b(0)|^2$ scales the scattering cross-section. Overall the expression describes the scattering by N uncorrelated identical particles.

In a concentrated system of discrete particles, the expression for the average intensity scattered takes the modified form:

$$< I(q) > \propto |b(0)|^2 P(q)S(q) \tag{3.8}$$

The additional term is known as the *static structure factor*. It represents the modification of intensity due to the spatial variation of the particles. In dilute systems S(q) = 1. Also, when q is high, $S(q) \rightarrow 1$. These features imply that the scattering intensity is dominated by the behaviour of P(q) when q is high, whereas S(q) dominates when $q \rightarrow 0$. The structure function is the feature particular interest, because its form and impact on <I(q)> has the potential to provide information about the fractal dimension.

The spatial correlations or structure in a concentrated system are usefully described by the *radial distribution function* g(r), which is related to the probability of finding the centre of any particle at a distance r from the centre of any given particle. Specifically, for N particles in volume V, the expression (N/V)g(r)dV represents the number of particles in volume element dV at a distance r from a given particle. In detail, the structure function is defined by:

$$S(q) = 1 + 4\pi \frac{N}{V} \int_0^\infty [g(r) - 1] r^2 \frac{\sin qr}{qr} dr \tag{3.9}$$

Fourier inversion of Equation (3.9) gives the radial distribution function:

$$g(r) = 1 + \frac{1}{2\pi^2} \frac{V}{N} \int_0^\infty [S(q) - 1] q^2 \frac{\sin qr}{qr} dq \tag{3.10}$$

In practice, the task is estimate S(q) from the scattering pattern and from this to discover the behaviour of g(r). When particles have a fractal structure, $S(q) \propto d^{-D}$. All being well, this is reflected in the <I(q)> dependence. An idealised scattering profile for fractal aggregates formed from spherical particles is illustrated in Figure 3.8. This can be split into three domains.

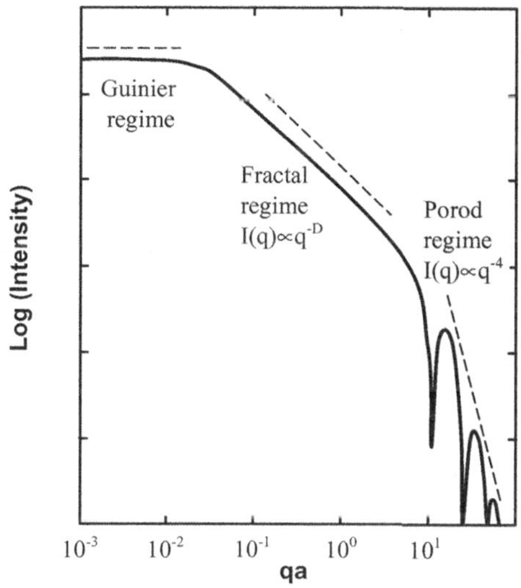

Figure 3.8. Scattering from aggregates formed from monodisperse spherical particles of radius a (adapted from Bushell *et al.*, 2002).

The Guinier region ($q \rightarrow 0$) reflects information gained from largest scale of the aggregates and is associated with 'edge effects' from the outer boundary. When q is large (Porod limit), the scattering intensity tends to follow the variation q^{-4} and is dominated by the form factor. When particles are monodisperse and spherical, one obtains a complex pattern described by Bessel functions, this being shown schematically in Figure 3.8. The intervening zone defined by:

$$1/R_G << q << 1/a \qquad (3.11)$$

is the one in which $\langle I(q) \rangle \propto q^{-D}$. By identifying this zone, the fractal dimension can be gained from the slope of the $\langle I(q) \rangle$ variation in a log-log plot. Inspection of Condition (3.11) shows that it refers to a region of space which is very much smaller than the overall size (indicated by the radius of gyration, R_G) and very much larger than the dimensions of the foundation particles.

When the form of particles and their spatial structure are known, the scattering can be predicted. However, the *inversion i.e.* the estimation of the structure from its scattering function cannot be solved uniquely (Glatter, 2004). To progress in a meaningful way, it is necessary to have a model, or series of models, against which to test experimental data. Pedersen (2004) provides a useful summary of some of the models that have been developed for colloids and polymer systems. Amongst these, Chen and Teixeira (1987) generated an expression for S(q) when the structure is fractal. The difficulty of interpretation is illustrated by the fact that a power law dependence $\langle I(q)\rangle \propto q^{-\beta}$ (with β constant) can arise in different ways. For example, Rasmussen (2001) notes that lamella particles, a polydisperse collection of spheres, and fractal aggregates *can each* result in a scattered intensity proportional to q^{-2} at small q. The study by Ramussen (focussing on the character of porous aluminium hydroxide) provides a good illustration of the analytical approach that needs to be adopted. Putting it forcefully, one cannot simply identify a regime in which $\langle I(q)\rangle \propto q^{-\beta}$ and to then interpret β as the fractal dimension. It requires far more supporting information and a good understanding of the scattering system.

Summary

From the review, the following aspects of radiation scattering are evident:

- At the present time, the only means of gaining insight into large scale floc structure is by small angle light scattering in conjunction with the RGD approach. In practice, this implies that the scattering particles must be smaller than 100 nm.
- Information about particle structures is favoured by conditions in which the scattering particles are identical. When the suspension is made up of aggregates of similar size and fractal structure, information about structure is constrained by the size domain specified by Condition (3.11). The constraint $1/R_G \ll q$ must be clearly understood, because it implies that one can only gain information about fractal dimensions at length scales *far removed* from the overall cluster size. If flocs have a multifractal structure, then the information yielded may refer to sub-structures. When flocs are multifractal, this will differ from estimates of D gained via sedimentation analysis. When the scattering particles are diverse in terms of size and form, the prospects for gaining meaningful information about particle structures are low.
- When using light as a radiation source, the RGD approach is best suited to structures of low fractal dimension (because of the potential problems posed by multiple-scattering from densely packed structures). This restriction matters less with X-rays and neutrons, because these forms of radiation are scattered weakly.
- Overall, considerable care must be exercised in the interpretation of scattering data.

3.1.3.4 Monitoring floc formation and breakage

In Section 3.1.2, attention was drawn to the difficulties of measuring floc sizes, because of ways in which floc properties interact with the processes of measurement. An alternative is the monitoring technique described in Gregory and Nelson (1986). This is based on the measurement of light transmitted through a flowing suspension. The instantaneous light intensity (I) can be written:

$$I(t) = I_o + I'(t) \tag{3.12}$$

The steady state value (I_o) is controlled by the average turbidity of the suspension, whereas the fluctuating component (I′) responds to random variations in the number and size of particles in the light beam. When a light sensor has a voltage proportional to the incipient light intensity, one can translate the ratio $\sqrt{(I')^2}/I_o$ into a corresponding voltage ratio, this being referred to as a *flocculation index* (FI) in Gregory (2004). The ratio always increases as aggregation occurs and decreases when aggregates are broken, the reason being that FI is roughly proportional to the square of the aggregation number (*i.e.* the mean number of particles per aggregate). Many studies, such as reported in Gregory (2004), show that FI is a sensitive indicator of the state of aggregation. Because turbidity is particularly sensitive to the presence of particles in the colloidal size range, the method provides insight into the small-scale processes that are taking place. This is of particular advantage when tracking the very early stages of aggregation. When aggregates are small, one avoids the problems of breakage within the measurement system. The main disadvantage of the method (which is far outweighed by the advantages) is that it is a semi-empirical index, which does not yield direct measurements of size. Gregory (2004) notes that the system is available in commercial form.

3.1.4 Strength

Background to the measurement of strength is given in Section 2.4. Apart from the study by Yeung and Pelton (1996), in which flocs were pulled apart mechanically (Figure 2.15), most estimates of strength are based on hydrodynamic rupture. Methods subdivide into those that provide a qualitative indication of strength, and those that are quantitative.

Here, we are concerned with the latter, these splitting roughly into two groups depending on whether the flow field is viscous or turbulent. The studies of Blaser (2000a, 2000b) provide an excellent example of the methodology that can be applied in viscous flows (Section 2.4.2). In turbulent flows, analysis has been severely constrained by Kolmogorov theory, which applies to breakage events at the extremes d << η or d >> η where η is the Kolmogorov length. In practice, most breakage events are in the intervening zone. It is here that the energy-based approach described in Bache (2004) circumvents this type of restriction as it applies to both laminar and turbulent flows. One also needs a model to represent floc strength, Equation (2.135) representing an example. Whatever model is used, attention must be paid to the definition of the parameters d and <ε>.

In a turbulent flow, it is unlikely that one will know the precise d-ε interaction, because of the way complex behaviour of ε (see Baldyga and Bourne, 1995) and arising from the inhomogeneous nature of flocs formed by random collisions. Thus, estimates of strength by this route are likely to be imprecise and subject to wide variation. It is generally accepted that the larger flocs are the most vulnerable to breakage, but any floc can be broken if it happens to encounter a regime in which there is sufficient kinetic energy on the scale of the floc to enable separation.

Many measures have been used to specify d_{max} shown in Equation (2.112). If distributions are self-similar with respect to G (Section 3.2.7.3), the mode of behaviour implicit in Equation (2.112) is independent of the choice of d_{max}. However, its definition does affect the magnitude of the strength scale. If strengths are to be compared, a consistent approach should be adopted. In the authors' studies (DHB), d_{max} has generally been defined by d_{V95} (the 95th percentile in the volume size distribution). The logic behind this choice is somewhat vague, but recognises the following: (i) rupture theory involves volume *e.g.* energy per unit volume; (ii) the volume measure (a distribution moment; Section 2.5.3) gives bias to the largest and weakest flocs; (iii) the 95-percentile is a convenient statistical descriptor of the upper limit.

The central problem surrounding <ε>, lies in the specification of its spatial distribution. In the studies reported in Glasgow and Hsu (1982) and Glasgow *et al.* (1986), $Fe(OH)_3$-kaolin flocs were allowed to settle into a horizontal turbulent jet in which <ε> had been estimated. The resulting data showed that strength (S) increased with size. This seems most unlikely, and one suspects that it was caused by errors in the estimation of <ε>.

A similar experimental approach was described in Bache *et al.* (1995), in which flocs were allowed to settle into a steady state turbulent field above an oscillating grid. The calibration of <ε> in this type of system is described in a Bache and Rasool (1996). A practical problem which was encountered, was the presence of large scale circulation loops which causes flocs to move along particular pathways, this also distorting the spatial distribution of <ε> and its calibration. Studies of this type depend on the observation of individual breakage events. These are most suited to suspensions in which the flocs are both large and opaque.

The approach is unsuitable for humic flocs, because of the combination of low settling velocities and optical translucence. In the case of Bache *et al.* (1995) and related studies, the association between strength and size could not be established in any meaningful way because of scatter in the data. Nevertheless, valuable information was gained (*e.g.* Bache *et al.*, 1999a). Another approach to resolving the problem of specifying $<\varepsilon>$, is to aim for a regime in which $<\varepsilon>$ is quasi-uniform. This can be approximated in a multigrid oscillating mixer, details of the turbulence regime being described in Bache and Rasool (2001b).

The vast majority of studies for measuring 'strength' focus on the d-G correlation in stirred beakers. These come in a large variety of forms, especially in the design of the impellers. For a simple impeller, the distribution of $< \varepsilon > / \bar{\varepsilon}$ follows the idealised spatial variation illustrated in Figure 3.9.

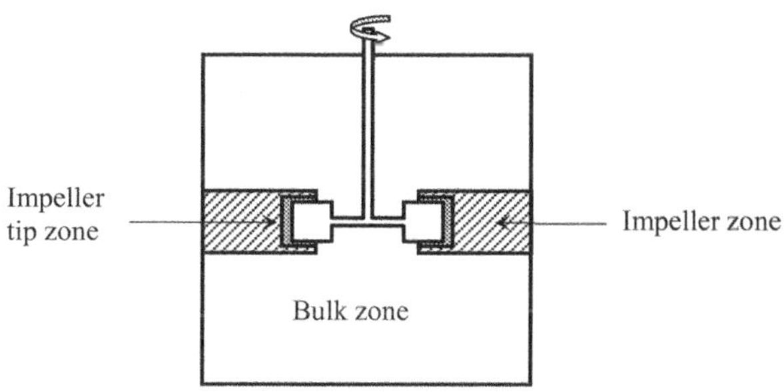

Figure 3.9. Idealised spatial variation of $< \varepsilon > / \bar{\varepsilon}$ based on a three zone model (Tomi and Bagster, 1978). The ratio $< \varepsilon > / \bar{\varepsilon}$ takes the values 0.25, 5.4 and 50 in the bulk zone, impeller zone and impeller tip zone respectively.

The fundamental problem with these systems is the lack of knowledge of $< \varepsilon > / \bar{\varepsilon}$. An exception is the two-litre Phipps and Bird square beaker stirred by a 20 x 75 mm simple blade impeller, whose calibration was reported in Stanley and Smith (1995). In this system the dependence of G on the rotation rate is also known (Cornwell and Bishop, 1983). As one would expect, Stanley and Smith's measurements showed that $< \varepsilon > / \bar{\varepsilon}$ has a continuous spatial variation throughout the mixer. In order to simplify the description, Bache (2004) analysed the published data and fitted a 'two-zone' model described in Okamoto *et al.* (1981). In this model, it is assumed that $<\varepsilon>$ is locally homogeneous within an inner zone of volume V_i (around the stirrer) in which $<\varepsilon> \equiv \varepsilon_i$ and is also homogeneous in an outer zone of volume V_o in which $<\varepsilon> \equiv \varepsilon_o$.

An energy balance within the vessel takes the form:

$$\rho_w V_i \varepsilon_i + \rho_w V_o \varepsilon_o = \rho_w V \bar{\varepsilon} \tag{3.13}$$

with:

$$V_i + V_o = V \tag{3.14}$$

where V is the total volume of the fluid. By analysing the data shown in Figures 10 and 11 of Stanley and Smith (1995), the following estimates were obtained: $\varepsilon_i / \bar{\varepsilon} \approx 4.5$, $\varepsilon_o / \bar{\varepsilon} \approx 0.5$ and $V_i/V \approx 0.13$. This provides a value for parameter f (= 0.5) in Equation (2.142). Once the d_{max} - $<\varepsilon>$ association is known, one can evaluate the various parameters used to characterise strength using the framework described in Section 2.4.5.

If a method is to be standardised for measuring floc strength, then there is considerable merit in basing this on the Phipps and Bird two-litre beaker in conjunction with the cited impeller design. All the essential parameters are known. Remote measurement systems should be used in cases where the flocs under scrutiny are likely to be ruptured in a sample stream. Thereafter the information can be analysed using the approach described in Bache (2004). The methodology has many drawbacks and weaknesses, but offers an approach that is physically-based, systematic and consistent.

3.1.5 Overview

- Considerable care must be exercised in the measurement of floc size, paying particular attention to the implications of shear within the measurement system.
- When floc are porous, there are considerable difficulties in assigning values to A′ and thereby to the specification of structure.
- Light scattering methods are fraught with difficulties in terms of providing information about the fractal dimension of flocs, particularly when composed of diverse materials.
- Particular attention needs to be paid to the specification of properties in the size range 1-50 μm. A major difficulty in this size range is the representation of flocs with a multifractal structure.
- There is a need to adopt a consistent approach in the measurement of floc strength. At the present time, the framework described in Bache (2004) can serve as a basis for progress.

3.2 IMPACTS ON FLOC PROPERTIES

3.2.1 Perspective and scope

Although there is a fairly good understanding of the factors that control coagulation, one cannot predict the effects of micro-constituents on the coagulation process - a reason why the jar test remains useful. In the case of flocs, the knowledge gap is greater. Floc properties stem from the founding materials in the raw water and the way these interact with materials which are added to facilitate treatment. In addition, one must take account of the array of kinetic factors involved in floc formation. Given the large number of variables involved, it becomes difficult to establish causal relationships. This is compounded by the difficulties faced in characterising flocs (Section 3.1). At full scale, a wealth of experience is balanced by a dearth of 'hard' information concerning the state of the solids phase.

Perhaps the most important parameter controlling structure formation is the colloidal charge (Section 1.2). If the charge is reduced, or its repulsive influence circumvented by bridging, structure formation can progress. When hydrolysing coagulants are used, the reaction timescales are such that structure formation begins with the coagulant itself. As hydrolysis progresses, the charge per metal ion reduces, as illustrated in Figure 1.24 or Table 1.5. This allows van der Waals forces to exert greater influence. If the rate of neutralisation is low, studies such as Bottero and Fiessinger (1989), have demonstrated both the existence and the evolution of colloidal structures as the degree of neutralisation increases. For rapid neutralisation (path I in Figure 1.29), the picture is less clear, because precipitate forms are both heterogeneous and amorphous.

In the case of aggregates formed from model suspensions under ideal conditions, there is general agreement that clusters formed by Brownian motion have a mass fractal dimension in the range of 1.7 to 2.1 (Section 3.2.7.2). In the orthokinetic domain, there are great disparities between studies, both in terms of the specification of floc properties and in the information generated. Part of this can be attributed to issues discussed in Section 3.1, and also arising from the difficulties of characterising multifractal structures. Much of the existing information is based on the study of model suspensions, and has tended to focus on larger floc sizes. There is a great shortage of structural information for aggregates within the 1-50 μm size domain (probably the most important size range in practical terms such as for DAF and in-depth filtration).

One of the distinguishing features of flocs is their water content, this affecting their strength, settlement properties and the dewaterability of the sludges which are formed (Section 3.3.7). Inspection of Equation (2.24) shows that increases in floc density are gained when the component solids have higher density (ρ_s) or when either or both the fractal dimension (D) and the packing factor (A′) have larger values or when flocs are smaller. Table 3.2 provides examples of features that affect the floc effective density or water content. The content is roughly correct, but there are always exceptions and mitigating circumstances that need to be taken into account when considering particular features. For example, when there is repulsion between particles the aggregation is said to be *reaction-limited* (RLA), the resultant aggregates being more compact than when particles are fully destabilised *e.g.* Firth (1976), Kim and Berg (1999). In contrast, data reported in Knocke *et al.* (1987) shows clearly that destabilisation by charge neutralisation (Sections 1.3.2.2 and 3.3.7.3) yields smaller and denser flocs than those associated with destabilisation by enmeshment in which a surface charge is likely to exist (Section 1.3.2.3). What appears to be a major contradiction can be partially explained if one takes account of the impact of the materials involved in the destabilisation process. Knocke *et al.* (1987) used aluminium sulphate as the coagulant, thus introducing [water-binding] amorphous precipitate (Section 3.2.2.3), this [quantitatively] having least impact during charge neutralisation. In Firth (1976), destabilisation was achieved by double layer compression (Section 1.3.2.1) - the zeta potential being controlled by an indifferent electrolyte.

Because of the many gaps in knowledge, Section 3.2 confines itself to themes in which there is a reasonable knowledge base. A major gap is the impact of water quality parameters, beyond the much-studied turbid suspensions and waters containing NOM. In places we address side issues that affect treatability. Finally, it must be acknowledged that much of the important foundation work in this area stems from the pioneering studies of Norihito Tambo of Hokkaido University in the late 1970's - this manifesting in a classic series of papers of which Tambo and Hozumi (1979) and Tambo and Watanabe (1979) are examples.

3.2.2 Coagulants

3.2.2.1 Background

Over the last two decades there has been a vast expansion in the variety of coagulants available for use in water treatment. Beside their role in destabilisation, these are also designed to address particular operating problems, such as control of pH or to enhance separation efficiency.

Table 3.2. Examples of factors which influence floc density. Symbols +/- indicate tendencies to increase (+) aggregate densities and (-) to cause higher water content.

Factor	Density impact	Comment	Section
Water components			
Hydrophilic materials	-		1.1.2
Non-isometric particles (e.g. mica , f ibres, filamentous algae)	-	Poor packing	
Micro-contaminants:			
alkalinity/hardness	+	Bonding by Ca and Mg (?)	3.3.7.1
DOC/NOM	-	*e.g.* via coagulant demand	3.3.7.3
Coagulant related			
Highly charge species	+	With good surface coverage	
Prehydrolysed	+		3.2.2.4
Hydrolysing			1.5
Destabilisation			
Near primary minimum	+		1.2.10
Secondary minimum	-	Particles held apart	1.2.10
Sweep floc	-	Hydrophylic materials	
Bridging	+/-	Depends on spacings	
Optimised coagulation	+		3.2.4
Mode of aggregation			
DLA	-	Neutral particles	3.2.7.2
RLA	+	Charged particles	3.2.7.2
Polymer flocculant	+(?)	Subject of debate	3.2.6.1
Ballasting agents, mineral turbidity	+		3.2.6.3
Rapid mixing	+	Multiple impacts	
Flocculation	+/-	Depends on G; + at high G	
Tapered flocculation	+		3.2.7.2

In low alkalinity waters it is often necessary to offset the acidity of the alum by lime addition. Such a requirement can be addressed in other ways, for example by using a coagulant which is less sensitive to pH (*e.g.* an iron salt), or using a coagulant such as PACl with higher basicity, thereby consuming less alkalinity. Another approach is to use a coagulant that has the capacity to deliver alkalinity *e.g.* an alkaline aluminium coagulant (Fettig *et al.*, 1996). Compounds are now being synthesized which combine the features of a coagulant and flocculant aid (Tang and Shi, 2002) with the potential to eliminate the separate addition of the latter. For example, activated silica (a special form of polysilicic acid) has long been used as an effective flocculant aid *e.g.* (Baylis, 1939; Masschelein, 1992). The usual practice has been to add activated silica separately from the traditional coagulant. Composites such as polyferric-silicate-sulphate (PFSS) and polyaluminium-silicate sulphate (PASS) avoid the need for dual dosing and appear to deliver greater aggregating power and separation efficiency compared with the traditional coagulants (Arnold-Smith *et al.*, (1992) and Gao *et al.*, (2000)).

While there is enormous experience of using the different coagulants in different situations, this is not matched by knowledge of the flocs. Beneath this is the question of the features that make one coagulant different from another. Some facets have been noted above. In many respects one cannot discuss this question in isolation from a particular water. Nevertheless, there are some traits that can be identified.

3.2.2.2 Mode of Destabilisation

A useful starting point is to consider the words of Tang and Shi (2002) "...the basic requisite characteristics of any coagulant or flocculant are the charge-neutralization and bridge-aggregation abilities for the removal of fine particles in raw water...".

Charge neutralisation is favoured by coagulants and their species which have a high charge and are sufficiently small to provide good surface coverage. This is why Al_{13} is particularly effective for coagulating particle suspensions. When a surface is covered in this manner, it can promote closer bridge binding between the particles at a lower dose. The latter feature was clearly demonstrated in Tang *et al.* (2004) in which the $AlCl_3$, $PACl_{2.2}$ (ligand ratio Al/OH=2.2) and Al_{13} were tested on a kaolin suspension. Al_{13} had the best coagulation efficiency (based on a lower residual turbidity at lower dose). The other feature of Al_{13} and $PACl_{2.2}$ (which contains a substantial fraction of Al_{13}) is that Al_{13} tends to aggregate to form clusters, branches and flocs.

The likely impact of coagulant charge on the density of flocs was reported by Jiang and Graham (1998) in a comparison of two forms of poly-alumino-iron sulphate (PAFS; with differing charge status) on a model coloured water. It was found that the PAFS with the higher charge generated the denser flocs. This implies better settleability (the source of the density information) and in all likelihood, better strength.

Bridging is enhanced by the presence of polymeric forms such as $Al_6(OH)_{12}^{6+}$ which plays an integral part in the "core-links" model (Section 1.5.3.3) or as a result of increasing the molecular size using compounds of the type noted above. With longer bridges, one should anticipate flocs with a greater water content, because the particles are held further apart. If this is accompanied by a reduction in the surface charge, there is the prospect of the flocs becoming more compact under the influence of van der Waals forces. However, for this to occur there must be accessible space to permit movement, together with sufficient free energy to meet associated demands. Bridging also plays a substantial role in destabilisation in sweep flocculation (Section 1.2.9).

When using traditional metal salts in this mode, or using pre-hydrolysed forms at a sufficiently high pH for hydrolysis to arise, the presence of bulk precipitate and its affinity for water (Section 1.5.7) suggests that that flocs are likely to have a greater water content - reducing their settleability and strength. When one has the combination of the bulk precipitate with raw water contaminants which are either hydrophilic or trap water mechanically (*e.g.* high molecular weight organic molecules), there is a potent recipe for flocs of high water content. In drawing attention to these features, one should not forget that the water content depends on factors such as rapid mixing (Section 3.2.7.1) and the manner of flocculation (Section 3.2.7.2).

3.2.2.3 Hydrolysing metal coagulants

Background to the coagulation behaviour of hydrolysing metal salts is provided in Sections 1.2 and 1.5; readers are also directed to the review of Duan and Gregory (2003). When using sweep flocculation, it must be recognised that hydrolysation and the precipitation of ferric salts are less sensitive to pH than alum. This is because ferric hydroxide is much less soluble than aluminium hydroxide and is precipitated over a wider range of pH (Figure 1.22). Hanson and Cleasby (1990) note that $FeCl_3$ is better for turbidity removal at the same molar dosage of Fe^{3+} and Al^{3+} - not withstanding the need to optimise pH. Similar observations are reported in Lindqvist et al. (2004) in the context of DOM removal. One significant factor is that at equivalent molar concentration, iron salts generate more solids by weight than aluminium (~ 2:1), a feature which can enhance floc size (see Gregory and Yukselen, 2002). The greater mass of iron has the disadvantage of increasing the sludge mass. Of the oxides and hydroxides, iron has the greater specific gravity (Table 3.3).

When precipitates form a major fraction of the solids mass (particularly in association with humic source materials), this is likely to enhance the settleability of iron flocs. At low temperature conditions, Duan and Gregory (2003) report that Fe(III) coagulants produce better turbidity and colour removal than alum.

However much depends on whether the pH has been properly adjusted. Even when pH adjustments are made, Gregory *et al.* (1996a) report difficulties associated with alum and the superiority of PACl in cold waters. Compared with prehydrolysed compounds, alum and ferric salts lack "refractory stability" (a term used by Tang and Luan (1996), seeming to refer to reactivity) making it more difficult to control their use. Hydrolysis is very rapid, and at the doses normally used, materials are rapidly converted into amorphous hydroxides.

Table 3.3. Examples of density of materials contributing to floc solids.

Material	Density ($kg\ m^{-3}$)	Reference
Humic substances	1570	Hossain and Bache (1991)
Synedra acus (diatom)	1100	Konno (1993)
Cryptosporidium parvum (oocyst)	1009	Dai and Boll (2006)
Giardia lamblia (cyst)	1013	Dai and Boll (2006)
Bentonite (≅ montmorillonite)	~2600	
Kaolinite	2610-2680	CRC (1980)
Montmorillonite	2000-3000	CRC (1980)
Silica sand (quartz)	2650	Perry (1950)
Activated carbon	1300-1500	Cheremisinoff & Ellerbusch (1978)
$CaCO_3$ (calcite)	2715-2940	CRC (1980)
$CaCO_3$ (aragonite)	2940-2950	CRC (1980)
$Fe_2O_3.\,xH_2O$	2440-3600	CRC (1980)
$Fe(OH)_3$	3400-3900	Perry (1950)
$Al_2O_3.3H_2O$ (gibbsite)	2420	Perry (1950)
$Al_2O_3.3H_2O$ (bayerite)	2530	CRC (1980)
$Al(OH)_3$(am) (pH 6)	2530	Hossain and Bache (1991)
$Al(OH)_3$	2420	CRC (1980)

When used for charge neutralisation, the relatively long times taken to deliver coagulant to colloid surfaces means that the species delivered to a target site is not necessarily in the most appropriate form for the task in hand. This reduces the efficiency of the process. Although charge neutralisation may be the central objective, it is largely inevitable that amorphous bulk precipitates will be present - a feature identified by Dempsey (1987) and discussed in Bache and Papavasilopoulos (2003). This is not necessarily a disadvantage because the precipitates are cohesive and serve as a useful adsorbent (Mazet *et al.* (1990), Cathalifaud *et al.*, 1993).

3.2.2.4 PACl

Polyaluminium chlorides (PACl) are widely used as coagulants and can be prepared with a range of ligand numbers (r = [OH]/[Al]) or basicities (% ≡ 100r/3). The higher the basicity, the less the alkalinity consumption (*i.e.* a reduced potential for altering pH). The basicity also affects the magnitude of the ionic charge (Table 1.4). At low basicity (say 33%) Edzwald and Tobiason

(1998) comment that these products contain mainly $Al(OH)^{2+}$, whereas Tang *et al.* (2004) associate the monomeric/oligomer fraction with Al_6. As the basicity increases, a greater fraction of the Al species resides in the form of Al_{13} (Tang and Luan, 1996). Beside their charge, the most distinctive characteristic feature of these materials is the stable structure of the pre-produced polymers, especially the Al_{13}. Tang and Luan state: "…after dosing these polymers can remain in optimal form refractory to hydrolysis before adsorption. On the surface the hydro polymers will undergo charge neutralisation and complexation with superficial OH groups. At a later time they can be further depronated and hydrolysed to become covered with aluminium hydroxide on the surface…" (see also Section 1.5.3.3). Using photometric dispersion analysis, Matsui *et al.* (1998) compared the flocculation kinetics of PACl and alum for a kaolin suspension. PACl worked faster than alum for corresponding Al. Reasons for this behaviour were not established. At the dose and pH values in use, both coagulants generate amorphous precipitate. However, PACl precipitates have a greater charge (Van Benscshoten and Edzwald, 1990a), this offering a greater probability of attachment with the clay colloids and stronger bonding compared with the alum precipitates.

Edzwald and Tobiason (1998) consider that higher basicity compounds (with greater Al_{13}) represent a good choice for the charge neutralisation of particles because of stronger attraction. They suggest that lower basicity products are better for the coagulation of NOM (presumably because optimum coagulation pH is lower and co-precipitation of NOM improves with lower pH). Gillberg (1994) demonstrated that the basicity of polyaluminium chlorides influenced the treatment of wastewaters and the quantity of sludge generated. Depending on the parameters under scrutiny *e.g.* turbidity or TOC, it was evident that optimal values of the OH/Al ratio could be identified for enhancing removal in jar tests. These examples highlight both the impact of basicity and the need to discover the most useful form for a particular application.

Attention has been drawn to the status of Al_{13} within PACl, aspects of its chemistry being covered in Section 1.5.3.3. Here, detail is added about its aggregation behaviour (in isolation from other materials) because this stamps its character on the floc. Basic features are summarised in Table 3.4. Below about pH 4, Al_{13} remains isolated by virtue of is strong surface charge. With increasing pH, it forms aggregates whose size and structure depend on the surface charge. Aggregates begin to form when the base ratio is about 2.3 (here the base ratio refers to OH/Al ion ratio of the solution. When the surface charge is high, the aggregates have a very open structure. When the pH increases, the surface charge decreases and the number of anchoring sites on each Al_{13} increases leading to a more dense structure (shown by the higher fractal index). The efficiency of Al_{13} for removing particles from raw water depends both on the structure and the surface charge. When the base ratio exceeds about 2.5, the system has the capacity to yield dense flocs that settle quickly. In this domain, the structure of the aggregates is less sensitive to pH.

Although this section concerns prehydrolysed forms of aluminium, it is worthwhile pointing out that the studies of Tchoubar *et al.* (1991) and Bottero *et al.* (1991) with respect to the partial hydrolysis of ferric chloride salt and ferric nitrate show the development of colloidal fractal forms which behave in much the same way as Al_{13} with respect to the solution OH/Me ratio. However, it is emphasised that the iron salts do not have a polymeric form equivalent of Al_{13}.

Table 3.4. Aggregation behaviour of Al_{13} behaviour (compiled from Bottero and Fiessinger, 1989). The base ratio refers to the composition of the solution. Parameter σ is an expression of the surface charge density, OH/Al refers to the solution and D to the fractal dimension.

OH/Al	pH	σ (H^+/nm^2)	D	Comment
<2.0	<4.06	1.55		Al_{13} units isolated
2.3	4.37			Turbidity increases
2.5	4.8	0.4	1.43	Open chain structures
2.6	6.0		1.86	Denser aggregates; flocculation threshold
>2.6	8.4	0		Isoelectric point; dense aggregates

3.2.2.5 Hybrid compounds

Beside the PACls, there are many other polyaluminium coagulants (PAXs) which include other additives. These offer differing mixes of basicity, silica and sulphur content. Added sulphate (typically <5%) enhances the precipitation of $Al(OH)_3$(am) and settling of floc. Silica improves both aggregation and settlement (Tang *et al.*, 1998) (see also Section 3.2.6.2). Edzwald *et al.* (2000) tested an array of these compounds on model raw waters with differing characteristics and solid-liquid separation processes. Some conclusions are noted, but it is more fruitful to read the original paper. Generally, best sedimentation performance was achieved using high basicity PAXs with sulphate. Silica-containing PAX could improve settlement. Good removals of NOM were found - higher basicity compounds offering a broader dosing range for good treatment. DAF was found to be very effective in treating low turbidity supplies of low and high alkalinity and supplies high in natural colour. Attention was also drawn to the problem of using direct filtration on a supply of low turbidity and low TOC (2.5-3.5 mg/l) *when* the NOM is composed of some aquatic humic matter. The coagulant demand caused by both the NOM and turbidity was such that the optimum coagulant dosages ($\geq$ 1.5 mg Al/l) were too high for direct filtration treatment because of excessive headlosses arising from the solids produced by coagulation. This applied to the full range of coagulants tested. NB. in practice, the limiting dose in direct filtration is dependent on factors such as floc strength, filtration rate, media size and depth (see Table 3.14 and Section 3.3.6).

Other papers which make extremely worthwhile reading include Tang *et al.* (1998) and Tang and Shi (2002). These papers give valuable insight into composition factors, mode of action and performance.

At least one proprietary product (Ca-PAX) is available with additional calcium, which may be helpful in coagulation of very soft waters. By including Ca in the product, it is possible to eliminate the somewhat problematic dosing system for calcium-containing substances, such as $Ca(OH)_2$ (Eikebrokk, 1996). In a comparison of aluminium sulphate, PAX 14 and Ca-PAX, (Eikebrokk, 1996) found that the calcium-rich product was effective over a broader range of pH for removal of humic substances - particularly where there was a high dosage requirement.

3.2.2.6 Organic coagulants

One of the attractions of using cationic polymers in place of metal ion coagulants is the potential to reduce the amount of sludge generated (Section 1.7) and thereby sludge treatment costs. Reductions in the solids load also has the benefit of extending filter run times where filtration is employed without clarification. Commonly used coagulant cationic polymers are PDADMAC and epiDM. Derivatives of tannin have also been used successfully. There is also considerable interest in the use of natural polymers such as chitin *e.g.* Divakaran and Pillai (2002) and those derived from the seeds of moringa olifeira *e.g.* Ndabigengesere *et al.* (1995).

When used as a primary coagulant, the main destabilisation mechanism is charge neutralisation, but with hydrophobic polymer-polymer interactions as a significant secondary influence in the overdose regime (Parazak *et al*, 1988). When promoting charge neutralisation, it has been found that polymers are most effective when their charge density (CD) is high. For the case of colour removal, this aspect is illustrated in Figure 3.10a. Maximum removal occurs when the charge density exceeds approximately 3 meq/g. This behavioural feature is also reflected in the correlation between the humic acid charge density and the charge density of the polymer (Figure 3.10b). When the CD exceeds 3 meq/g there is a one-to-one charge interaction, so that charge neutralisation is brought about when sufficient polymer has been added, irrespective of its CD value. When the polymer CD has a lower value, there is a non-stoichiometric interaction between the anionic sites of the humic substances and the cationic charges of the low charge polyelectrolytes.

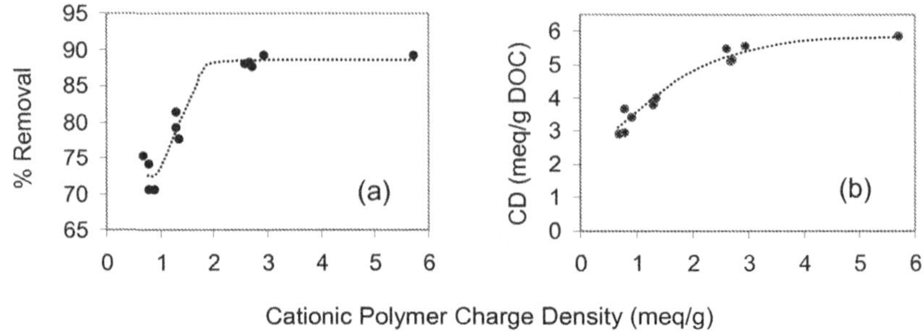

Figure 3.10. (a) Degree of removal of humic acid for different cationic polyelectrolytes as a function of their charge density (CD) (b) Average apparent charge density of humic acid at pH 7 as a function of polyelectrolyte charge density (based on Kam and Gregory, 2001).

Polyelectrolytes with metal ion salts

When waters contain NOM, cationic polymers work best when there are particles present in the water (Bolto *et al.*, 2001). Where waters lack turbidity, particles can be added by adding clay, metal oxides *e.g.* ferric oxides or via the use of a hydrolysing metal coagulants such as alum. The solids have the potential to adsorb NOM; the role of the polymer is to coagulate the organic–laden particles. For one water studied, a 67% reduction in the alum dose was possible by adding 1 mg/l of cationic poly-DADMAC, to give an equal or better performance (based on reduction of UV absorbers) than alum used alone at the optimum dose.

Beside the role of adsorption noted above, Ammary and Cleasby (1995) suggest that the contribution of the inorganic hydroxide precipitate to the total suspended solids concentration improves flocculation kinetics, while the cationic polymer produces larger and stronger flocs.

From studies of the coagulation of a silica suspension using both hydrolysed Al and polyelectrolyte coagulants, Letterman and Sricharoenchaikit (1982) reported a simple trade-off relationship between the amounts of Al and polyelectrolyte required to reach the state of charge neutralisation. Their study also showed that the highest fractional turbidity removals were obtained when charge neutralisation was accomplished using polyelectrolyte and no aluminium was present. This was attributed to the differences in the floc settlement rate, the polyelectrolyte yielding flocs of higher density than those from hydrolysed Al.

On the question of the relative amounts of alum and polymer, Gregory (1991) and Hall (1997) refer to a "technical optimum" substitution of alum by the polymer (when DOC/NOM is present). For greater substitution, the coagulation process becomes dominated by the polymer such that the colour reduction becomes increasingly worse and unacceptable whereas the residual turbidity improves. In this context, it should be noted that Lindqvist *et al.* (2004) found that organic polyelectrolytes are not as efficient as inorganic metal salts in DOM removal. A comparative study of coagulant type (including poly-DADMAC) on humic floc structure, notional strength and settleability is reported in Jefferson *et al.* (2004).

Practical issues

When using a cationic polymer for coagulating low turbidity waters at low temperature, Ammary and Cleasby (1995) found that the low temperature had a detrimental effect on the flocculation kinetics (as occurs with alum). Given the substantial differences between these coagulants, it suggests that the impact on flocculation is probably a manifestation of the effects of temperature on viscosity, rather than arising from changes in the system chemistry.

The order of dosing metal ion coagulant and cationic polymer has not been identified as an issue, although Ammary and Cleasby (1995) reported that simultaneous additions produced the strongest flocs. Of greater significance is the relative proportion and optimal dosage of the two. When chemical handling facilities are limited, operators have produced the desired effect by blending the two products prior to dosing. At least one blend of ferro-aluminium sulphate and a polyDADMAC has been commercially available in the UK as an approved product.

There are several reservations for the routine use of cationic polymers as a coagulant:

(i) Experience shows that the quality of the treated water in terms of TOC and trihalomethanes (THM) precursors may not be satisfactory, especially when the cationic polymer dose is maximised.

(ii) Although methods exist for measuring polymer residuals in water (*e.g.* Keenan *et al.*, (1998), Bache (2002)) these do not match the ease (or sensitivity) of monitoring and measuring the residuals of iron and aluminium.

(iii) A related concern is that the final TOC of the treated water will not be wholly natural (*i.e.* NOM); a proportion of the final TOC will be residual polymer, which is synthetic and therefore less acceptable than NOM.

3.2.3 Size-density relationship

The size-density relationship (Equation 2.24) conveys crucial information about floc structure. Its functional dependence is normally gained via sedimentation analysis. The interdependence is not readily extrapolated beyond the size range associated with the original measurement. Its form is sensitive to the coagulation conditions and the forming materials. Studies such as Tambo and Watanabe (1979) (based on sedimentation behaviour) indicate that the size-density function is relatively insensitive to shear. However, Bache and Rasool (2006) found that the size-density measured by this approach could not account for the overall solids mass within a stirred suspension. Further there were indications that the prevailing dependence was sensitive to shear. The appraisal provided in Section 3.1.3.2 indicates that estimates of the packing factor (A′) is less reliable than that of the mass fractal dimension. With knowledge of the size-density relationship it is relatively straightforward to gain estimates of the average solids volume concentration within a floc. This in turn provides insight into floc permeability and its rheology.

Tambo and Watanabe (1979) explored the impact of a variety of factors encountered in treatment on the size-density function. Using both coloured water and a clay suspension, this focussed on factors such as dose, pH, alkalinity, coagulant type, flocculent aids and mixing intensity together with others. In the context of clay-aluminium flocs, it was shown that the floc density function was largely insensitive to pH, alkalinity and the presence of flocculant aids. Of the various findings, perhaps the most significant was the impact of the *ALT ratio*, this referring to the metal ion concentration divided by the suspended solid concentration in the treated water.

For example, a dose of 5 mg Al/l in a suspension with SS 250 mg/l has an ALT ratio of 1:50. By varying the dose and suspended solids concentration, different values of ALT were generated. In some cases, the mix was adjusted to yield the same ALT ratio. The resulting trends are displayed in Figure 3.11. These show that the highest density flocs (ALT = 1:100 and across all sizes) arise when the ALT ratio is least, and the lowest density when the ALT is highest (1:10). In physical terms, the ALT is more or less scaling the fraction of the floc solids occurring in the form of the bulk precipitate, this having the ability to trap water (Section 1.5.7). Thus when the ALT fraction is highest (*i.e.* most precipitate), the lowest effective density is a direct consequence of the water binding capabilities of the precipitate (see also Section 3.3.7). Trends shown in Figure 3.11 were cast in the form of Equation (2.24) using data cited in Tambo and Watanabe (1979), together with the values d_o = 3.5 μm and ρ_s=2650 kg m^{-3}, this forming the basis of Figure 3.12.

With some allowance for scatter, the trends indicate that both A′ and D are uniquely defined by the ALT value. This implies that when a subset of flocs is characterised by an ALT value (*i.e.* any trend in Figure 3.11), they share the same fractal structure (specified by A′ and D) and solids composition (specified by ALT) irrespective of size. Of the two parameters, A′ has the greatest impact on the size-density relationship. Bearing in mind the comments about the water binding capability of the bulk precipitate, this seems to be the factor which drives the association between ALT and A′; a curious feature of the plots in Figure 3.12 is the appearance of turning points in the D and A′ trends at

ALT = 0.02 - perhaps associated with charge neutralisation, whereas sweep floc dominates at greater ALT values.

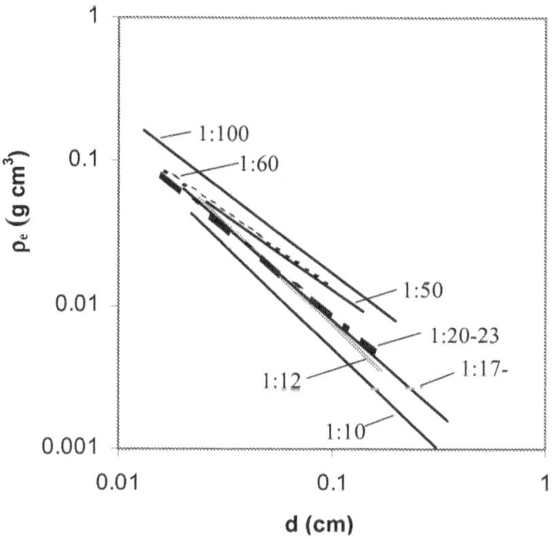

Figure 3.11. Influence of ALT ratio on size-density plots (data from Tambo and Watanabe, 1979).

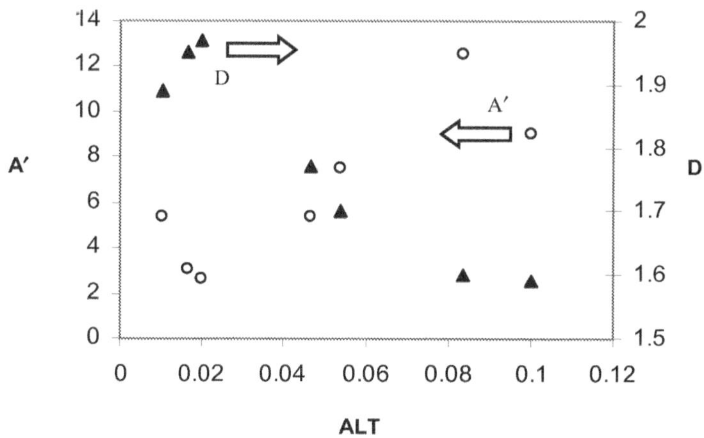

Figure 3.12. Dependence of packing factor and fractal dimension on the ALT ratio for trends shown in Figure 3.11. Data apply to both alum and PACl as sources of Al (based on data cited in Tambo and Watanabe, 1979).

Figure 3.13 shows a set of trends mapping the size-density function for flocs sampled from a series of water works treating low turbidity coloured waters.

Details of the water quality characteristics and coagulation practices are summarised in Bache *et al.* (2000). Trends were fitted by Equations (2.24), values of A′ and D being reported in Bache (2004). Apart from the plot for Wybersley, the remainder have similar slopes (1.0 < D < 1.2), but differing intercept values *i.e.* differing values of A′. The plots fall into two groups depending on the presence or absence of a polymer floc aid. The effect of the floc aid is to increase A′, and thereby strength (see Equation 2.133). The trend for Wybersley is of interest because it approaches the size-density relationships associated with polymer addition - even though there was no polymer in use. At the time of sampling, the water at Wybersley was a blend (75% soft upland and 25% calcareous ground water - whereas all the other waters were soft upland). The enhanced density might be attributed to the hardness - an aspect that requires investigation. Overall, the set demonstrates the role of A′ (rather than D) in type-casting the different waters and coagulation practices.

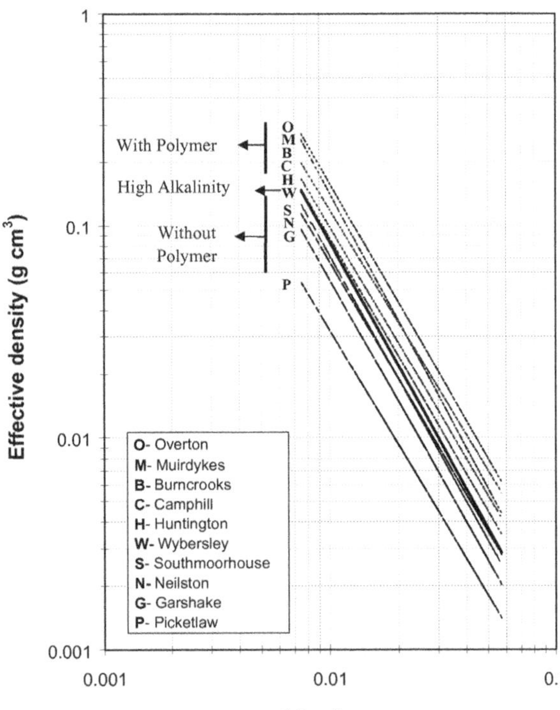

Figure 3.13. Summary plot of size-density distributions at different WTWs (Bache *et al.,* 2000).

A similar feature was evident in a previous study of a low turbidity coloured water reported in Bache and Hossain, (1991). In this earlier study, information was gained about the response of the size-density function to the coagulant dose. This focussed on a dose regime both above and below the optimum dose (5 mg Al/l) determined by jar tests. It was found that the size-density relationship passed though a maximum as the dose was increased (a feature which was insensitive to pH; see Figure 3.15). Using a reference floc size, this pattern of behaviour is illustrated in Figure 3.14. Beyond the maximum, the reduction in density can probably be attributed to excess precipitate and water binding. Below the maximum, the cause of increasing density is unknown, but must centre on the interaction between the humic substance and the coagulant.

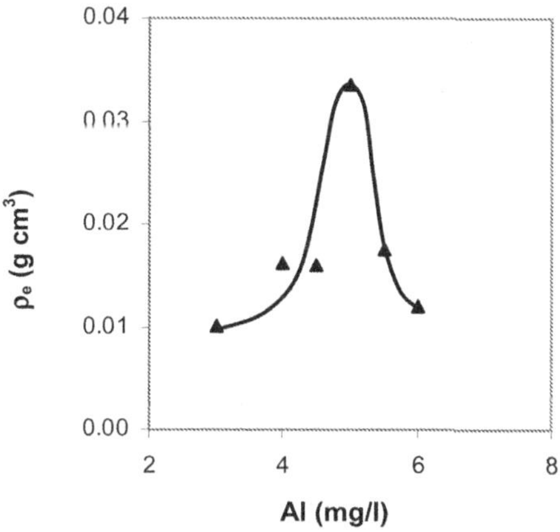

Figure 3.14. Response of effective density to dose for 200 μm flocs gained from a low turbidity coloured water at pH 4.75 (based on data in Bache and Hossain, 1991).

3.2.4 Optimised coagulation conditions

Although the jar test is well established for guiding optimum coagulation conditions in terms of the system chemistry, it provides little insight into floc properties. Initially one wants to know how floc properties respond to the coagulation conditions, and further, to understand how their properties are likely to affect solid-liquid separation (Section 3.3). The first theme has been addressed in Bache and Hossain (1991) and Hossain and Bache (1991) in the context of a low turbidity coloured water and Bache *et al.* (1995) concerning a turbid suspension (destabilised by sweep floc). In both studies, the coagulant was aluminium sulphate, and optimum coagulation conditions were based on the supernatant water quality. Thereafter a suite of parameters was examined to characterise the floc; this included the size-density relationship, average floc size, settleability and strength. Of the two studies, the latter was better founded in terms of instrumentation and methodology. The important outcome was the demonstration that floc properties displayed optimal features, with maxima occurring at much the same dose and pH as the optima identified by jar testing. In both studies, it was evident that overdosing caused a deterioration in floc properties, a feature that can be attributed to excess quantities of bulk precipitate and the water that it traps.

Coloured waters

Figure 3.15 illustrates the association between optimal floc properties and data from jar test analysis for the case of the coloured water. From jar tests, the optimum coagulation conditions were judged to be about 5 mg Al/l and pH 4.8 at 20°C. This lies within the cross-hatched regime delineated by the intersection of the zones associated with the supernatant residuals Al $\leq$ 0.2 mg/l and colour $\leq$ 5 Hazen. The zone of minimum colour is seen to be controlled by pH, this coinciding with the conditions in which the solids release is maximised (see Figure 1.33b). The trend delineating the maximum floc density (ρ_e) was based on a reference size (similar to plot shown in Figure 3.14). This trend is seen to be controlled by dose rather than pH. Its striking feature is that it passes through the cross-hatched zone showing where both residuals are minimised. The plot also shows trends delineating maxima in a characteristic floc size (d_{50}) and v_{10} (characterising settleability) together with a region within which floc strength was judged to be maximised. Viewed as a whole, one can see a general consistency between optimum floc properties and the findings of jar test analysis. In terms of Figure 3.15, the most significant features controlling floc properties appear to be the pH at which the solids release is maximised, and the dose at which the floc density is maximised. When these occur, it maximises the floc solids volume concentration (ϕ) which in turn controls strength, size and settleabilty.

Turbid suspensions

The correspondence between floc properties and the optimum coagulation conditions (by jar testing) for a turbid suspension is described in Bache *et al.* (1995). The investigation focussed on a 250 mg/l suspension of rice starch granules coagulated with aluminium sulphate. Flocs were characterised in terms of the size, density and strength and settleability, pertinent data at the optimum pH being summarised in Table 3.5.

The salient feature of the data shown in Table 3.5 is the general consistency between optima in floc properties and those identified by jar testing. Trends in the strength and size were characterised by distinct maxima with reduced values in conditions of under-dosing and overdosing. The impact of floc density and size was represented by a weighted average velocity across the size distribution. It was found that <v_{mt}> diminished progressively with increasing dose, reducing from around 0.6 mm s^{-1} at 3 mg Al/l to about 0.2 mm s^{-1} when the dose was 10 mg Al/l. The diminution of size and strength in the overdose regime is worthy of comment. The sensitivity to pH is illustrated in Figure 3.16.

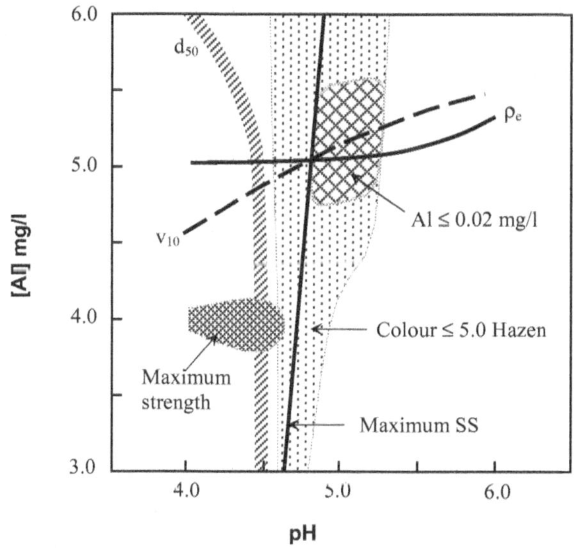

Figure 3.15. Dependence of residual colour and aluminium concentration on the Al dose and pH. Also included is the zone of maximum strength, together with trends in the maxima of size (d_{50}), velocity (v_{10}), effective density (ρ_e) and the maximum in the dry solids production (Bache and Hossain, 1991).

Table 3.5. Critical doses in a turbid suspension at pH 6.3 and 20°C (Bache *et al.* 1995).

Parameters	**Critical dose Al (mg/L)**
Jar tests	
95% turbidity reduction	4.8
minimum residual	3.5
Floc parameters	
maximum vertical transport <v_{mt}>	~ 2.5
maximum floc density (d>40μm)	2.5 - 4
maximum strength	4 - 6
maximum floc size (d_{V95})	~ 5

Implications

The authors are unaware of other comparisons of this type. On the basis of these studies, it appears that optima identified in jar tests are broadly consistent with the conditions at which optimised floc properties exist. This conclusion is somewhat tentative and *does not* imply that the jar test automatically identifies the conditions favouring treatment.

Enlarging on the above, jar tests can be considered as a reasonably effective means of identifying conditions for treatability, but not the outcome at full-scale. The principal reason is that the separation processes taking place in a jar test are radically different from those encountered during treatment. In addition, both floc structure and size distributions alter as the suspension passes through a treatment works. Nevertheless, if the removal of the principal contaminants at full scale differs greatly from that indicated by jar tests, then it points to inadequacies in the application of the chosen chemistry.

The interpretation of jar test data depends on the objectives. If one is concerned only with turbidity removal (as has been the case in parts of the world for waters with relatively low colour) then a lower alum dose (as might be associated with charge neutralisation) as indicated in Table 3.5 may well be appropriate. However, when removal of NOM is also a requirement, then a higher alum dose (producing sweep floc conditions) may well be necessary. In such case the optimum dose is associated with the diminishing return in increasing coagulant dose in improving and balancing of the removal of turbidity, residual coagulant and NOM.

Beside the jar test one should be aware of the wide array of test methods and systems for supporting operational control of coagulation, clarifiers and filtration; see for example: Hall (1997), AWWA (1992) and Dentel (1991).

3.2.5 Temperature

Temperature influences coagulation, flocs and floc formation in many ways - the impacts on system chemistry and viscosity being the most significant. At low temperatures, operational problems are often reported (particularly when using alum) with slow growth, small flocs and poor quality residuals (*e.g.* Morris and Knocke, 1984). Some of these effects can be ameliorated by: altering the pH; using higher coagulant doses; or by adding floc aids. Prehydrolysed coagulants such as PACl have gained favour for use at low temperatures because of their superior performance. In the case of coloured water, the colour precipitates better (less soluble) such that less coagulant appears to be needed.

3.2.5.1 System chemistry

Chemistry is specific to a particular coagulant and the associated water quality factors. In turbid suspensions, the system chemistry is largely unaffected by the presence of the colloids, whereas in the case of coloured waters, coordination reactions occur between the coagulant and humic materials, this affecting the reaction kinetics. In spite of these differences, some behavioural features are shared. For example, studies by Kang and Cleasby (1995), Dempsey (1987), Van Benschoten and Edzwald (1990a) and others have shown that Fe(III), Al(III) and PACl all display a solubility–pH dependence which shifts to the alkaline side as the temperature is reduced. There is also a lowering of the theoretical minimum solubility as the temperature is reduced, but this does not always occur in practice

The implication of the above is that the optimum coagulation pH shifts to a higher value at low temperatures, a feature borne out in studies such as Dolejs (1984), Van Benschoten and Edzwald (1990a,1990b), Bache *et al.* (1996) and Gregory *et al.* (1996). This is to be expected because the shift in overall solubility stems from changes of the charged hydrolytic components in the dissolved phase (see Edzwald and Tobiason, 1998). Similar features are also evident in the behaviour of the solid phase. For example Van Benschoten and Edzwald (1990a) showed that alum precipitates maintain their positive charge up to about pH 9 at 4°C whereas the isoelectric point (zero electrophoretic mobility) at 25°C was pH 7. The charge on PACl colloids also falls with increasing pH, but is less affected by temperature.

On the basis of studies of ferric sulphate and alum, Hanson and Cleasby (1990) and Van Benschoten and Edwald (1990a) found that solubility kinetics at low and high temperature were nearly identical if the pOH was kept constant. When the pOH is constant, the thought is that this helps to maintain a consistent Al/OH ratio and thus consistent conditions for the formation of the hydroxides (Section 1.5). The relationship between pH and pOH is determined by the relationship:

$$pH + pOH = pK_w \tag{3.15}$$

in which pK_w is an equilibrium constant defined by:

$$pK_w = -6.0875 + 4470.99/T + 0.01706\,T \tag{3.16}$$

with T as the absolute temperature. If a pH *e.g.* 6.3 is identified as the optimum at a particular temperature *e.g.* 20 °C, then the corresponding pOH is 7.87. If we then assume that pOH 7.87 is the target pOH, then the corresponding pH at other temperatures can be calculated using Equations (3.15) and (3.16). Examples of this are summarised in Table 3.6. Taking the maximum and minimum pH values shown this corresponds to a pH shift of 0.034°C^{-1}. In practice, the coagulation kinetics tend to show a smaller sensitivity to temperature. For a turbid suspension coagulated with alum, Bache *et al.* (1996) reported changes close to 0.021 C^{-1}. In the case of humic suspensions Dolejs (1984) reports changes of 0.01 C^{-1}, whereas Van Benschoten and Edzwald (1990b) indicate a more sensitive dependence. From extensive testing of a range of commonly used coagulants on in waters of different character waters in the range 2 – 25°C Gregory *et al.* (1996) report shifts of about 0.5 pH unit (≡ 0.022 °C^{-1}) for alum, but less for iron salts.

Table 3.6. Dependence of pH on temperature at a constant pOH (from Bache *et al.*, 1996). Also included is temperature dependence of dynamic viscosity (μ) and kinematic viscosity (ν) for water.

T (°C)	pK_w	pH	pOH	$\mu \times 10^3$ (Pa s)	$\nu \times 10^6$ ($m^2 s^{-1}$)
0	14.94	7.03	7.87	1.792	1.792
5	14.73	6.86	7.87	1.519	1.519
10	14.53	6.66	7.87	1.308	1.308
15	14.34	6.47	7.87	1.140	1.141
20	14.17	6.30	7.87	1.005	1.007
25	13.99	6.12	7.87	0.894	0.896
30	13.83	5.96	7.87	0.801	0.804
35	13.68	5.81	7.87	0.723	0.727
40	13.53	5.66	7.87	0.656	0.661

3.2.5.2 Flocs and floc formation

Changes in viscosity affect the rate of flocculation and hence the flocculation timescale. Data in Table 3.6 show the dependence of viscosity on temperature. The combination of Equations (1.14) and (1.12) shows that the flocculation timescale is proportional to the dynamic viscosity in the perikinetic domain, whereas in the orthokinetic domain its impact is less obvious. In the latter domain, Equation (1.23) indicates that the flocculation timescale is controlled by G which is defined by Equation (1.39). In turbulent flow, the timescale follows the approximate dependence $\tau \propto \nu^{1/2}$; this arises from the G dependence shown in Equation (1.39), but is modified by Reynolds number dependencies associated with the power input (P). Viscosity also effects floc rupture, its impact encapsulated in Equations (2.137) and (2.138).

While these effects exist, it is generally accepted that the system chemistry is the more important. This aspect is illustrated in Figure 3.16, which shows the sensitivity of floc development to pH. This refers to a 250 mg/l suspension of rice starch coagulated with alum. At 20 C, jar tests indicated optimum dose as 5 mg Al/l at pH 6.3. The striking feature of the plots is the behaviour either side of the optimum pH. When pH ≥ 6.3 the plots are similar, in contrast when the pH is below the optimum, the rate of growth is slower, this having the consequence of limiting the floc sizes attained in the short-term *i.e.* smaller flocs.

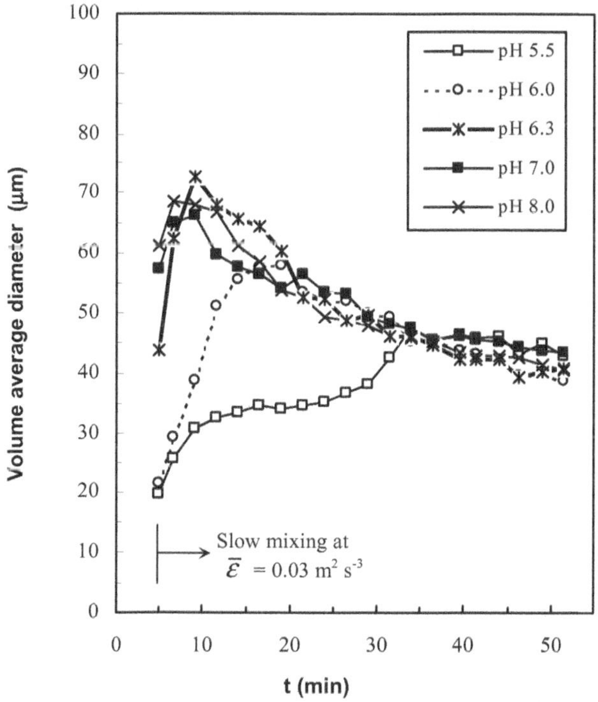

Figure 3.16. Showing dependence of volume average diameter on pH for a suspension of rice starch at 20 °C coagulated with alum at 5 mg Al/l (Bache *et al.*, 1996).

Of particular interest is the behaviour at pH 6.0 *i.e.* only 0.3 units below the optimum, showing that relatively small changes in pH *(below the optimum)* can have a significant impact on floc development. The reasons for this behaviour are unclear, some candidate factors being shown in Table 3.7. In the context of Equation (1.21), the mass of precipitate affects the rate of flocculation, whereas the zeta potential and strength (through its impact on size) each affect the trapping efficiency (α). However, the ways that these interact is unclear.

Commenting on poor turbidity removal at low temperatures, Morris and Knocke (1984) linked this to floc characteristics on the basis of flocs being smaller than those formed at 'normal' temperature. In reaching this conclusion, Morris and Knocke took no account of pH. Figure 3.16 shows that the behavioural feature described by Morris and Knocke can arise when the pH is below its optimum at the same temperature.

Table 3.7. Data associated with Figure 3.16. Bulk precipitate refers to dry mass within the suspension. Terms σ_{80} defines strength percentile at which 80% of measured flocs had strength less than cited value. The zeta potential refers to coated rice starch granules (Johnson, 1998).

pH	Bulk precipitate $Al(OH)_3(s)$ (mg/l)	Strength (σ_{80}) (N m^{-2})	Zeta potential (mV)
5.5	12.9 ± 0.3	2.5	27
6.0		3.6	27.5
6.3	13.3 ± 0.4	1.2	25
7.0	19.8 ± 0.4	0.4	21.5
8.0		0.8	13

In contrast, and focussing on the related aspect of floc development, Camp *et al.* (1940) stated "...changes in temperature have no measurable effect on the time of floc formation, if coagulation takes place at the optimum pH...". This aspect was scrutinised in Bache *et al.* (1996). The trends shown in Figure 3.17 illustrate floc development and floc size under conditions in which the pH was optimised (through jar testing) at three different temperatures. Note the range of temperatures examined and the associated pH values. During flocculation, adjustments were made to the mixing regime to ensure that flocs were subjected to the same $\bar{\varepsilon}$ value (essentially eliminating viscous influence on hydrodynamic stresses encountered by the flocs). In spite of the large temperature range, the upper sizes are similar and behave in a same way with respect to time. There is no evidence of the behaviour associated with trends pH < 6.3 shown in Figure 3.16. The evidence appears to support the views of Camp *et al.* (1940). Detailed analysis of flocculation timescales reported in Matsui *et al.* (1998) provide some support for the pOH route (discussed above) for adjusting the coagulation chemistry.

If pH is not adjusted to take account of shifts in the solubility pattern with temperature, then the symptoms described by Morris and Knocke may well occur. Here it is worth noting that at low temperatures (with pH well below its optimum value) the combination of increased viscosity and smaller sizes reduces the settling speed of flocs. Thus if solid-liquid separation depends on settlement (as in the jar test), the separation process becomes much slower and less efficient. This forms part of the explanation for the 'poor turbidity removal' at low temperatures when pH control has been neglected. It is not the complete story, because experiences show that there has to be some adjustment of dose as well as pH at low temperatures (Dolejs (1984), Gregory *et al.* 1996).

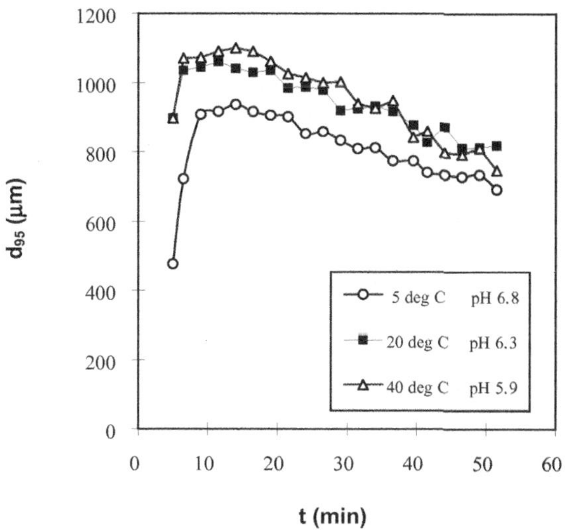

Figure 3.17. Development of the upper floc sizes (d_{95}) as a function of temperature at pH optima at an Al dose of 5 mg/l (Bache *et al.*, 1996).

3.2.6 Flocculant aids

Flocculation aids refer to additives that assist the formation of larger and denser flocs - features that facilitate their separation in clarifiers and filters. Polymer flocculants are widely used, others including activated silica and bentonite.

3.2.6.1 Polymer flocculants

When an aluminium or iron salt has been used as a primary coagulant, the polyelectrolyte has capacity to increase the floc size, density and strength, features which enhance settleability. Lockyear *et al.* (1983) draw attention to the considerable effect of flocculant aids on floc size, but remark that they have little effect on final water quality (colour and turbidity). Nevertheless, the study of Lindqvist *et al.* (2004) illustrates the benefit of adding cationic polyacrylamide (as a flocculant aid) to the removal of the smallest MW fraction of DOM (< 1kDa) - particularly when used in conjunction with ferric salts). Dosing polyelectrolyte to increase floc size and strength is often useful when water is coldest in order to counteract the effects of water viscosity in sedimentation and filtration. Without adjustments in pH or shear, floc formation in very cold water (especially with aluminium sulphate) can be very slow and dosing polyelectrolyte can help to compensate for this (Gregory *et al.*, 1996a); this also applies to the situation in which pH has been adjusted.

Following coagulation, particles/microflocs will generally be in a state of low surface charge whose sign depends on the mode of destabilisation, dosage and pH. Even when a destabilised particle is 'neutral' there may be available sites of different sign where polymer segments can absorb as a result of electrical interactions or hydrophobic attraction. Polymer flocculants are most effective when they have a high molecular weight, bridging being their most important attribute (see also Section 1.2.8).

There are a great many polyelectrolytes available and approved for use in water treatment as flocculants. Products approved are listed only according to basic chemical type and approved dosage limit. However, the potential user must take account of the charge type, charge density and molecular weight.

Choice of flocculant aid

Because of the multitude of parameters involved (raw water type, primary coagulation, polymer characteristics, costs and tasks ahead, *etc.*), the choice of a polyelectrolyte often requires an extensive testing programme to find the most appropriate product for a given task *e.g.* Dentel *et al.* (1988, 1989). In general, jar testing provides a rapid screening procedure for flocculants. This phase of testing can be refined by adopting procedures which provide more detailed information on settlement, flotation or filtration, depending on the separation processes ahead. Ultimately the choice can be backed by pilot plant testing or testing at full scale.

The literature review presented in Gregory and Miller (1994) identified a number of methods for selecting the most appropriate product, but it gave little indication of any clear guidelines as to which type of polymer works best in particular circumstances. For most polyelectrolyte investigations, it was concluded that the choice was site specific.

Bratby (1981) compared flocculant aids when used in conjunction with alum. The water used was generally of low colour and turbidity, and very low alkalinity (below 6 $mgCaCO_3/l$). Initial tests involved ranking the polyelectrolytes in terms of floc formation time, floc size and order of settling. Of fourteen products tested, the non-ionic forms were slightly better than an anionic product, with the cationic products being markedly inferior. At lower doses, the anionic product proved to be the best giving good overall performance.

Lockyear *et al.* (1983) mentioned cationic products (as flocculant aids) being likely to perform better with highly coloured water. In contrast, Amirtharajah and Tambo (1991) suggested that cationic products were better for clay particle coagulation and ineffective for coloured water (when used alone). Singley (1972) suggested that lower molecular weight cationic polymers would be more effective than similar weight anionic or non-ionic products for inter-particle bridging.

Gregory *et al.* (1996a) assessed polymers of differing charge type on typical waters found in the UK (Table 3.8) using a suite of test methods. In all cases aluminium sulphate was the primary coagulant, with the dose and pH optimised. The relative performance of the polyelectrolytes was ranked using a scoring system. The outcome of the assessment is summarised in Table 3.9.

Table 3.8. Quality of raw waters. Notation: high alkalinity river water (W); high alkalinity eutrophic water (X); low alkalinity, low coloured water (Y); low alkalinity, coloured water (Z) (Gregory *et al.*, 1996a).

Determinand	**Water**			
	W	**X**	**Y**	**Z**
Conductivity (μS/cm)	615	591	40	78
Alkalinity (mg $CaCO_3$/l)	278	223	13	30
pH	8.2	8.6	5.8	6.6
Orthophosphate (mg P/l)	0.44	0.64	0.02	0.06
Total aluminium (mg Al/l)	0.07	0.03	0.10	0.17
Total iron (mg Fe/l)	0.13	0.07	0.07	0.22
Turbidity (NTU)	16	3.1	0.67	4.6
True colour (°Hazen)	10.73	3.85	17.33	37.95
Apparent colour (°Hazen)	52.53	24.20	26.68	76.18
Apparent uV abs. at 254 nm	0.720	0.282	0.606	1.104
Chlorophyll_a (μg/l)	11.9	93.3	2.7	3.65
Coagulant dose selected (mg Al/l)	3	4	1	1.5
Coagulation pH selected	6.65	6.5	6.2	5.5

Table 3.9. Efficacy of polymer flocculants with respect to water type shown in Table 3.8 (Gregory *et al.*, 1996a).

Polyelectrolyte	**Behaviour**
Strongly anionic	Effective for all waters tested except organic rich water (Z)
Non-ionic*	Most effective for soft and most coloured water (Z); least effective for other waters tested
Strongly cationic	Most effective for soft and most coloured water (Z); moderately effective for the other waters tested

* common usage refer to ampholytic polymers as non-ionic

Table 3.10. Polymer usage at full scale.

Planning factors

Alternatives: Before embarking on polymer dosing, check the apparent need for polymer, in that operational problems might be approached in other ways *e.g.* changes in filter media or improved backwashing arrangements.

Mud-binding: *Mud-binding* refers to a condition which can become apparent at the end of a filter run, visual evidence including the existence of cracks on the surface of a filter bed and the presence of abnormal clumps of congealed media which resist breakage during backwashing. These conditions are exacerbated by excess polymer. Other pointers include a deterioration of filter performance (rapid headloss and breakthrough). When this occurs, the polymer dose should be reduced and the dosing strategy reviewed. It is very important to avoid this type of problem. If mud-binding has occurred, and the filter bed cannot be effectively washed by combined air-water, then the contaminated filter media must be removed and replaced.

Polymer selection and target dose: Select polymer and effective dose on the basis of tests representative of the eventual application; data from jar tests and pilot plants must be treated with caution. If the effective dose appears to be greater than about 50% of the maximum approved dose, then the selection procedure should be reviewed.

Monomers: If a polymer is used for washwater and sludge treatment, and supernatant is recycled, then the monomer content must be taken into account in limiting the maximum dose of polymer for clarification (or *vice versa*).

Changes in water quality: Raw water quality can change profoundly on a seasonal basis, and it may be appropriate to change the polymer type: *e.g.* when TOC concentration is high, a cationic/non-ionic polymer might be most effective; when turbidity is high, an anionic/non-ionic polymer might be the best choice.

Dosing strategy

Solution preparation: Manufacturer's instructions for preparation of the polymer solution should be followed. Use as low a solution concentration as is reasonable. Ensure that all polymer is in solution prior to dosing, and that the desired concentrate concentration has been attained. When make-up water is close to freezing, it may be necessary to heat the water. Use carrier water if appropriate to facilitate dispersion into the coagulated water. Avoid dosing at too small a relative volume (polymer solution: raw/coagulated water flow rate).

Initial dosing: When first introducing a polymer into a full-scale plant, caution should be exercised: start with about 20% of the indicated effective dose. Allow a full cycle of filtration for all filters between increases in polymer dose (also about 20% of expected effective dose) until an adequate response is achieved. If the effective/optimum dose appears to be greater than 50% of the approved maximum dose then review the situation, e.g. addressing polymer suitability, preparation, dispersion, delay time after dosing of coagulant (NB should be as long as possible).

Dose adjustments: The polymer dose should be adjusted as the raw water quality and coagulant dose vary. The expected polymer dose should be roughly proportional to the coagulant dose. As with other chemicals, the applied dose should be proportional to the receiving flow.

Optimum dose

The optimum dose of a flocculant can only be examined in the context of the process(es) ahead so that they can be driven harder in terms of increased throughput and/or superior treated water quality. At this point it suffices to state that there is often a practical end-point which limits the dose. For example in filters overdosing causes excessive headloss and in the extreme can damage the filter beds. In the case of a floc blanket sedimentation (Section 3.3.4), there is little point in applying a polyelectrolyte or a greater dose of polyelectrolyte if the blanket floc concentration is greater than about 18% (Gregory *et al.*, 1999). Lockyear *et al.* (1983) state that the optimum dose "...will often be found to be generally in proportion to the dose of primary coagulant (iron or aluminium salt)...". A similar feature (tied to polymer adsorption on hydroxides) has been reported in Bache and Papavasilopoulos (2003) in the context of polymer conditioning for dewatering (see also Section 3.3.7.5).

Optimising use

Identification and good control of the optimum coagulant and polymer doses and coagulation pH do not guarantee good plant performance, though they are a necessary part of this objective. Other necessary requirements include: (i) rapid dispersion of the chemicals immediately they are dosed (Watts, 1994); (ii) appropriate order of dosing (especially of acid or alkali and coagulant (Gregory et al., 1996a); (iii) time delay between dosing of each chemical, especially between coagulant and polyelectrolyte (Yadav and West, 1975; Gregory *et al.*, 1996a) and (iv) time between dosing of last chemical and entry to the treatment process (Yadav and West, 1975). Some general advice on the use of flocculants at full scale is summarised in Table 3.10.

Impact on flocs

Although there is extensive experience of the use of flocculant aids in practice, their impact tends to be judged by process performance, with little information provided about the state of the associated floc. In the research literature, there is a considerable body of knowledge concerning polymer-particle interactions and the role of polymers in destabilisation. In contrast, there is relatively little information about the ways in which polymers function as a floc aid *i.e.* when used jointly with a primary coagulant. The conventional wisdom is that they bind preformed flocs together by bridging. When used alongside hydrolysing metal coagulants, there is evidence of interactions with the hydroxide, particularly in the context of dewatering (Section 3.3.7.5).

Many aspects of floc character are reflected in the state of the size-density relationship. In the context of alumino-humic flocs, inspection of the trends shown in Figure 3.13 indicates that when a polymer is present, they have a higher effective density and thereby a greater solids volume concentration within the floc. In terms of a structural model *e.g.* Equation (2.24), it appears that the effect of the polymer is to increase parameter A′ (subject to caveats noted in Section 3.1.3.2), rather than the fractal dimension. Tambo and Watanabe (1979) examined a variety of floc aids (organic polymer, sodium alginate and activated silica) on the character of clay-aluminium flocs. They reported that floc structures were strengthened (indicated by the response to shear), but that the size-density function *was largely unaffected i.e.* maintaining the same values of A′ and D.

In some respects, Tambo and Watanabe's observation also applies to the alumino-humic flocs referred to in Figure 3.14, in that associated increases in strength when a polymer was present could not be fully explained by increases in A′ (Bache, 2004). In the context of Equation (2.132), the strength depends on an energy term $\beta\varphi$ as well as A′. Aside from these components, the crucial information scaling strength is embodied in the factors S_o and S_o' shown in Equations (2.132) and (2.135). Values of S_o for different floc types are shown in Table 2.2. It is evident that the magnitude of S_o is greater when a polymer was present. The impact of these factors on floc size can be estimated using Equation (2.137) and relative changes in size by Equation (2.144). In the text below Equation (2.144), an example is provided showing that the enhancement of S_o by polymer addition is consistent with a doubling of the average floc size. Increases in size have the knock-on effect of increasing settlement velocities, and often improving separation efficiency.

3.2.6.2 Activated silica

The use of activated silica in water treatment was prompted by Baylis (1939), who found that silica in natural waters enhanced flocculation when using alum, and to a lesser extent in the case of ferric coagulants. The term "activated silica" implies a suspension containing negatively charged colloidal particles, obtained by reaction of sodium silicate with an activator, usually an acid. The resulting sol (as used in water treatment) is a polymer of partially neutralised sodium metasilicate (details in Masschelein, 1992). The size of the sol is in the range of 0.1 to 1 μm and is built from base particles of about 5 nm in size. The dry material has a BET surface area of about 250 m^2/g. Packham and Saunders (1966) found that when a small amount of activated silica was added in advance of aluminium sulphate, significant enhancement of coagulation occurred between pH 5.5 and 6.7. When used as a flocculation aid, activated silica is usually added at the stage of pinfloc formation (Masschelein, 1992). Claimed benefits include a broader pH range for effective coagulation, more rapid settlement, better performance at low water temperatures and enhanced colour reduction.

The impact on settlement appears to be the result of increased strength (through its impact on size) rather than increased density. As noted above, Tambo and Watanabe (1979) found that the size-density relationship for clay-aluminium flocs was largely unaffected by flocculant aids, their study including activated silica. However the impact on floc size is quite marked as illustrated in Figure 3.18 - the maximum floc size increasing with the activated silica concentration. The links with strength are expressed by Equation (2.144), with further detail Bache (2004).

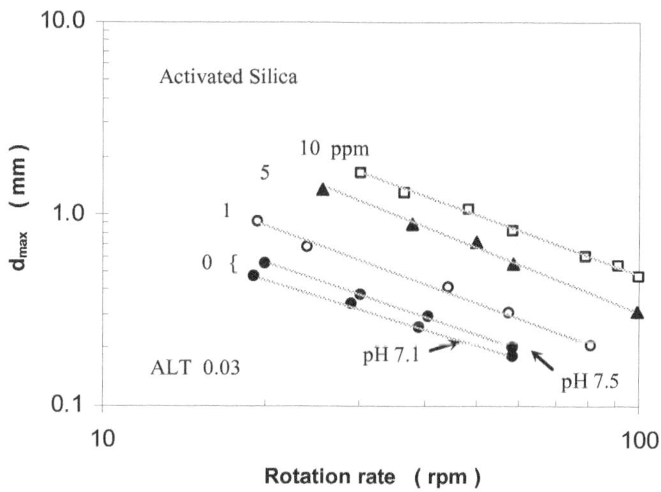

Figure 3.18. Relationship between d_{max} and impeller rotation rate (controlling degree of shear) for clay-aluminium flocs with variable activated silica dosage (based on Tambo and Hozumi, 1979).

With regard to naturally occurring silica found in water (concentrations commonly in the range 1-30 mg/l as SiO_2), this is likely to exist in the monomeric form of silicic acid $Si(OH)_4$ (Duan and Gregory, 1996). Duan and Gregory examined the impact of silicic acid in this concentration range on the flocculation of a clay suspensions coagulated with alum. Depending on the combination of concentration and pH, it was found that flocculation could be both enhanced or inhibited depending on the combination of concentration and pH. This appeared to be tied to the impact of silicic acid on the surface charge of aluminium precipitates, and thereby on the precipitation kinetics, but the precise mechanisms were unclear (Duan and Gregory, 1998).

3.2.6.3 Ballasting

Resumé

Ballasting is the use of materials that increase the weight (and possibly size) of flocs that result in faster settlement. Materials used for ballasting can be inert or active, in that they chemically or physically react in some way with impurities. Microsand (<200 μm) is used in proprietary processes – the sand being recovered for recycling through use of hydrocyclones. Powdered activated carbon dosed before clarification acts as a ballasting agent.

Floc produced by silt or clay-bearing water generally settles faster than the floc produced by coagulating water containing little mineral turbidity. Consequently, mineral turbidity added purposely to increase floc density can be useful. Bentonite is the usual choice, and fly-ash has been considered in Eastern Europe. The advantages of ballasting floc also arise when using powdered activated carbon (Standen *et al.*, 1995) or when precipitation softening is carried out in association with iron coagulation.

A process based on microsand was developed in Hungary in the 1950s and 1960s, known variously as Simtafier (Webster *et al.*, 1977) and Cycloflocs (Sibony, 1981). The process has been developed further by French companies to be known as Densadag (Degrémont (1991); see also Dauthuille, 1992) and Actiflo (de Dianous *et al.*, 1990) systems which include inclined settling. The Fluorapide system (Sibony, 1981) combines floc blanket settling and sand ballasting with inclined settling.

Richard and Capon (1980) refer to an "inorganic filler" (mainly sand) as maintained in fluidisation. Coagulation-flocculation results from the contact with this filler - which permits the density of the floc formed to be increased. They also regarded the microsand acting as a filler.

Typically in the Actiflo process, the raw water is first dosed with coagulant. After about 2 min mixing the microsand is dosed together with about 30-70% of the polymer. After a further 2 min the residual (70-30%) polymer is dosed and about 6 min allowed for maturation prior to lamella settling. Typical microsand doses are 3 mg/l, in range 2-4 mg/l.

As with ordinary coagulation, Actiflo has low efficiency for waters with a low SUVA value (refers to Specific UV Absorbance, *e.g.* Edzwald, 1993). High sand dose adsorbs excess polymer and reduces filter blinding.

Desjardins *et al.* (2002) studied ballasted flocculation in the laboratory by jar tests. The results showed that coagulant dose and coagulation pH have the greatest effect on settled water quality. By contrast, filterability was more sensitive to coagulant and microsand doses. Filterability was also dependent on the ratio of microsand to polymer dose and improved when this increased.

A related process is Sirofloc, in which powdered magnetite is used as an absorbant surface and its magnetic quality allows it to be 'flocculated' and recovered for recycling. Chemically activated magnetite is mixed with the raw water and then treated with cationic polymer prior to passing through a magnetic field and then fast settlement, Gregory *et al.* (1988).

Bentonite

Bentonite is an alumino-silicate clay which has been used to improve the coagulation of waters containing insufficient suspended particles, especially when colour removal is the main concern. Beside its role as a weighting agent, it has a considerable adsorptive capacity which can benefit treatment. Its role as a flocculation aid has been reviewed in Olin and Peterson (1937) and Patel and Shah (1978).

Its impact on floc settlement was scrutinised in Al Farisi (1988), some key findings reported in Bache *et al.* (1995). The Al Farisi study focussed on practices at a water works in Scotland treating low turbidity coloured waters with aluminium as the coagulant. At the time of the study, the clarifiers suffered from wind disturbance, bentonite being added to improve the stability of the floc blanket. The bentonite doses used in practice were 3–12 mg/l. Using the same water and replicating the coagulation practices, Al Farisi examined the impact of bentonite dose on the size-density function. As the bentonite dose increased, it was found that the floc density passed through a maximum located in the interval 3–7 mg/l *i.e.* similar to the doses used in practice. For a 100 μm floc, the effective density (ρ_e) increased by about 200 kg m^{-3} in the optimum dosage range (see Bache *et al.*, 1995).

3.2.7 Shear

Flocs and shear are rather like Castor and Pollux (heavenly twins of classical mythology) - shear representing an omnipresence which moulds the state of the floc. Experiments of the type described in Logan and Kilps (1995) indicate that aggregate properties are influenced by the structure of the shear regime encountered during their formation. A glance at the Subject Index is perhaps the best means of appreciating the full extent of the interaction. Of particular note are sections of the text dealing with flocculation, rheology and strength, these embracing the basic elements of growth and rupture. When considering impacts of shear, it is useful to draw a distinction between regimes in which flocs exist as discrete entities and those in which the solids form a continuous network (see Figure 2.12c). Here, the focus is on the former, the intention being to address some themes involving the impact of G on dilute suspensions.

3.2.7.1 Rapid mixing

The prime purpose of rapid mixing is to ensure that chemicals are quickly dispersed throughout the flow. The factors discussed in Section 1.4.3 indicate that mixing timescales are reduced by increasing the power input per unit volume *i.e.* by increasing the G value. There is also the issue of delivering a reactive chemical such as a hydrolysing coagulant in a form which is fit for purpose, this being tied to the mode of destabilisation. A related benefit of rapid mixing is that it helps to 'kickstart' floc growth through the growth phase bounded by 1 μm (the limit of perikinetic growth) at its lower end to sizes ~ η (the Kolmogorov length) at the upper end. When the value of G is high, not only is the range narrowed (because η is smaller), but the flocculation rate (proportional to G) is also enhanced. As well as affecting growth, rapid mixing also influences aggregate structure.

Rapid mix and destabilisation

When using hydrolysing coagulants, the impact of rapid mix depends on the mode of destabilisation. In the case of adsorption and charge neutralisation (Section 1.3.2.2), Amirtharajah and Mills (1982) state "…it is imperative that the coagulants be dispersed in raw water as rapidly as possible (in less than 0.1 s) so that the hydrolysis products that develop in 0.01 to 1 s will be adsorbed and hence cause destabilisation of the colloid…". The principal reason for recommending contact times of this order is that shorter timescales favour the existence of more highly charged species within the hydrolytic chain shown by the sequence in 'Equation' (1.47) or in Table 1.5. If one can take advantage of the higher charge, it has the prospect of achieving neutralisation at a lower dose, thus making more efficient use of the applied chemical. Experimental evidence supporting this view is provided in Klute and Amitharajah (1991, p237). At the practical level, it is difficult to deliver contact times in the order of 0.1s – particularly in back-mix reactors (stirred by an impellor). However, in-line pipe mixers are advantageous for this application (see Klute and Amirtharajah, 1991).

In the case of sweep floc (Section 1.3.2.3), Amirtharajah and Mills (1982) suggest that the time of hydroxide formation is 1 to 7 s, suggesting that high intensity mixing is "...not as crucial as in adsorption-destabilization…". Their study showed that the residual turbidity was relatively insensitive to G values in the range 300 to 16,000 s^{-1}. A different view emerges when considering the strength of the flocs. Figure 3.19 shows a comparison of the floc strength at extremes of mixing practice, 'poor practice' being simulated by adding a colloid suspension 5 min after the addition of aluminium sulphate. The deterioration in strength was considered to be caused by low charged bulk precipitate interfering with electrostatic bridging (Section 1.2.9). A similar theme is described in Zhanpeng and Yuntao (2006, Table 3), emphasising the advantage of maintaining good dosing practices when using sweep floc.

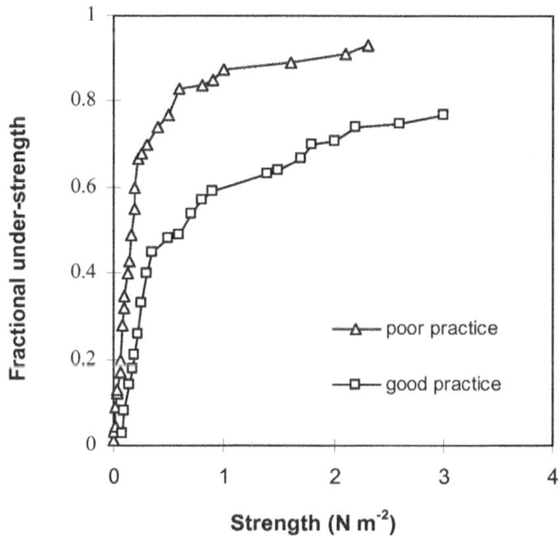

Figure 3.19. Sensitivity of strength status to rapid mixing practices for a Minusil suspension (Bache *et al.*, 1999a).

Optimisation

Beyond the dispersion stage, rapid mixing is just a manifestation of orthokinetic flocculation at high G. As such, one must expect the process to have similar control parameters to those described in Sections 1.3.3.3 *i.e.* involving the mix of concentration, velocity gradient and time. Perhaps the clearest exposition of this feature has been the investigation reported in Letterman *et al.* (1973). This was based on the behaviour of an activated carbon suspension coagulated with alum. The basic approach was to fix a concentration mix (suspension and aluminium dose) and G and to examine the subsequent impact of mixing duration on residual turbidity. The latter was measured in the supernatant after a standardised sequence involving rapid mix, flocculation at low G and settlement. As the mixing time increased, it was found that turbidity-time dependence followed a characteristic pattern in which turbidity showed a rapid decrease with time (presumably in response to aggregate formation), passed through a minimum, and then increased. The Gt combination associated with minimum turbidity (an optimum) was found to comply with a dependence of the form $Gt\phi^n$ = constant with n > 0. The significant feature was the existence of limiting Gt combinations, overmixing (Gt above optimum) causing a deterioration in supernatant quality. Similar findings are reported in Tambo and François (1991, p265). Remarks in the latter seem to imply that floc sizes are maximised at the optimum Gt, a feature which would partially explain the nature of the turbidity-time trends reported in Letterman et *al.* (1973).

Impact on structures

There is a school of thought, *e.g.* François and Van Haute (1985), that flocs formed during rapid mixing represent the basic building blocks (*flocculi*) of larger scale floc structures. During rapid mixing, flocs are subject to the highest level of shear that they are likely to encounter in their history. In these circumstances, the surviving flocs will be small and dense. Their size can be roughly estimated by the association $d_{max} \sim \eta_{rapid}$, where η_{rapid} refers to the Kolmogorov length scale associated with rapid mixing. A more detailed view of size would need to include strength factors (Sections 2.4.4 and 2.4.5). From changes in the slope of the d_{max}-G dependence, such as reported in François and Van Haute (1985), it appears that small flocs in equilibrium with high levels of shear have a greater fractal dimension than the larger flocs existing at lower G. In the case of ferric hydroxide flocs, Wiesner (1992) estimated these subunits to have a fractal dimension of 2.35, a value appearing to comply with a range 2.1-2.6 reported in Li and Ganczarczyk (1989) for aggregates of similar scale.

3.2.7.2 Restructuring

Typical symptoms of restructuring are illustrated in Figure 3.17, which shows a progressive reduction in floc size with time beyond the maximum. Under constant shear, reductions in size are a strong indicator that densification is taking place, this being brought about by structural deformations (*e.g.* bending of particulate chains and bond rotations) and by successive break-up and re-aggregation. Restructuring can occur in different phases of flocculation. For example, Chakraborti *et al.* (2003) showed that restructuring took place in the early stages of floc formation. In a similar vein, Jung *et al.* (1996) found that the fractal dimension of iron hydroxide flocs increased from 2.25 to 2.52 as hydrolysation progressed. In the same study (using small angle light scattering), the authors found that the fractal dimension increased from 2.52 to 2.71 as shear was increased. This was attributed to restructuring, though it may also point to the existence of a multilevel structure. Bache and Rasool (2006) found that changes in structure during mixing was one of many ways by which a size distribution adjusts itself to the prevailing shear conditions.

When clusters are formed by random collisions, the structure attained is not necessarily in its most stable state. Insight into some of the factors involved is provided by Aubert and Cannell (1986), who studied the aggregation of silica colloids ($\approx$ 22 nm diameter) in Brownian conditions. Aggregation was controlled by altering the ionic strength of the solution - this affecting the height of the energy barrier and thereby the magnitude of the collision efficiency factor (α) (Section 1.3.3.2). When α was high (associated with *diffusion limited aggregation* or DLA), Aubert and Cannell found that clusters (a few μm in size) were characterised by a fractal dimension of 1.75 ± 0.05. However, when α was low (with slower or *reaction limited aggregation* (RLA) caused by the existence of an energy barrier), clusters had D = 2.08 ± 0.05 *i.e.* denser packing. Depending on the particular ionic conditions and the passage of time, it was found that the mass fractal dimension of the clusters formed by DLA always reverted to 2.08 ± 0.05. Key features arising from the study include: (i) the sensitivity of the fractal dimension to α; (ii) the existence of restructuring - indicating that the bonds formed by DLA are not fully reversible (*i.e.* unable to attain their original form); (iii) an indication that structures formed by DLA can be metastable, thus possessing a tendency to convert to more stable forms and (iv) the value D $\approx$ 2.1 for clusters formed in a model suspension by perikinetic flocculation.

The reversibility or non-reversibility of structural forms may be demonstrated by increasing the shear rate on preformed flocs (causing breakage) and restoring the original shear rate (referred to as "cyclic shear" in Spicer *et al.*, 1998). These features are illustrated in Figure 3.20. If the original size is restored, the system is reversible. Commenting on this phenomenon, Duan and Gregory (2003) pointed out that clay particles destabilised using cationic polyelectrolytes (involving charge neutralisation and electrostatic patch effects) show almost complete re-formation after breakage, whereas irreversibility "seems to be specific to sweep coagulation". Clearly the experiences of Aubert and Cannell (1986) indicate that irreversibility is not specific to sweep coagulation. In the case of the latter, irreversibility is probably tied to the impacts of shear on electrostatic bridging (Section 1.2.9). When flocs are formed under good mixing conditions, much of their strength can be attributed to effects of electrostatic bridging. If some of these bonds are broken and reform, there will be a tendency for lower charged bulk precipitate to get into the intervening space. This will cause a reduction in the strength of electrostatic attractive forces as a result of increase separation distance (N.B. parameter r in Equation 1.9; see also Bache *et al.*, 1999a). This is similar to the phenomenon illustrated in Figure 3.19.

Spicer *et al.* (1998) note that when bonds are unable to reform to their previous extent, it reduces the collision efficiency factor. Simulation models such as Clark and Flora (1991) and Meakin (1988), indicate that when α is reduced, its lower value can be offset by colliding aggregates inter-penetrating and creating more bonding opportunities. When aggregates coalesce in these circumstances, the structure will have a denser packing than cases in which α is higher (requiring less inter-penetration to establish coalescence).

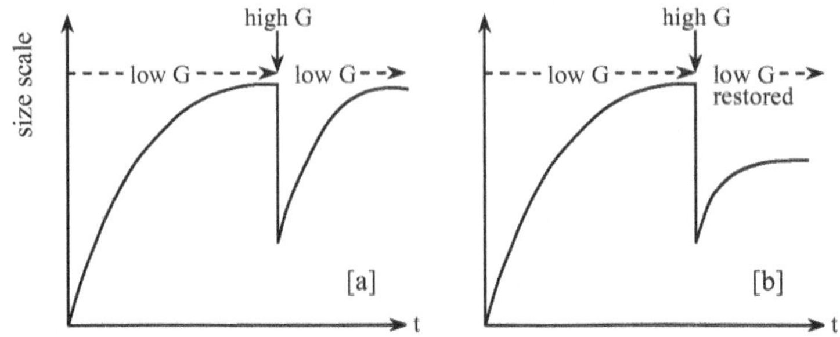

Figure 3.20. Evolution of size in a flocculator under cyclic-shearing. In case (a), the equilibrium size is restored following intense shearing, implying bond reformation. In case (b) there is a reduction in the equilibrium size following intense shear; this shows irreversibility and indicates that restructuring has taken place.

When polystyrene-aluminium flocs were subjected to cyclic shearing, Spicer *et al.* (1998) showed that reductions in the mean size were accompanied by increases in both the average floc density and the mass fractal dimension. An interesting development of their study, was the demonstration that phased tapered-shear flocculation in which G was gradually reduced from 300 s^{-1} to 50 s^{-1}, resulted in flocs that were denser than those produced at constant shear (50 s^{-1}) or by cycled-shearing. It was suggested that the denser flocs produced by tapered-shear regime were more amenable to separation by processes such as settlement and centrifuging, as originally demonstrated in TeKippe and Ham (1971). Tambo and Wang (1993) indicate that the prolonged agitation of "random flocs" (*i.e.* built by random collisions) can lead to pellet formation (clusters with high fractal dimension, say $D > 2.5$), though these are normally formed by one-by-one attachment of the primary particles to existing clusters.

Selomulya *et al.* (2003) showed that flocs formed from larger primary particles (810 nm) were more susceptible to breakage than those formed from smaller primary particles (380 nm). However, the principal outcome of this study was an attempt to model the size changes associated with restructuring using a population balance model. With the help of semi-empirical fitting functions, their model was reasonably successful in reproducing the observed trends. The exercise provided some insight into the underlying kinetics.

In addition to "mechanical changes" affecting structural forms, structural changes can also be driven by changes in the system chemistry. This is well illustrated by the effects of hydrolysis on floc properties through time when using traditional coagulants. As hydrolysis progresses (*e.g.* following the model in Figure 1.24) reductions in the surface charge of the polymeric forms alters the state of bonding. The consequences of ageing on clay-aluminium flocs were described in François (1987). These manifest as changes in size, density and strength, factors which can have a major impact on the properties of stored hydroxide sludges. For example, the data shown in Figure 3.21 shows how the yield shear stress of a hydroxide sludge reduces with time, a symptom of structural changes.

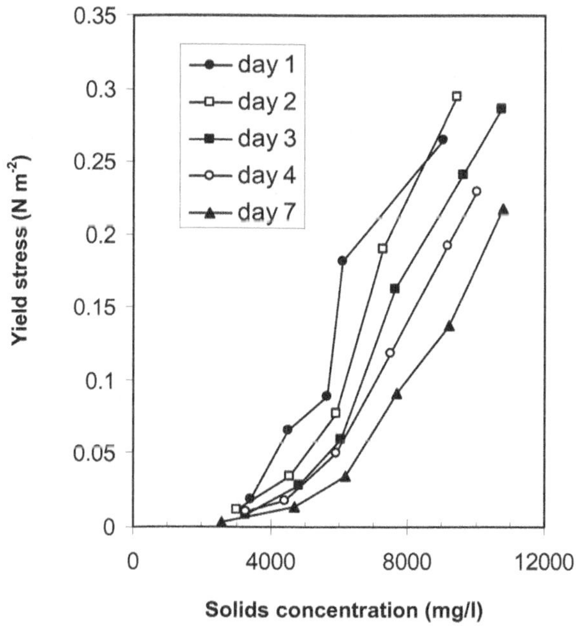

Figure 3.21. Changes in yield stress with time for an hydroxide sludge formed from alumino-humic flocs (unpublished data, DHB).

3.2.7.3 Impacts on distributions

Resumé

In continuous flow processes, such as found in water treatment works, distributions of floc size are usually in a steady state or slowly varying. Knowledge of distributions is desirable, because many separation processes discriminate on the basis of particle size. The success of separation is normally judged in terms of the mass removal efficiency, implying that it is advantageous to know the association between the distribution of mass and size. In some processes, *e.g.* floc blanket clarifiers (Section 3.3.4), one requires knowledge of the overall floc volume. A framework for providing such information from the number-size distribution (Ψ) is described in Section 2.5. For clarity, some basic concepts are reiterated.

Superficially, the impact of shear on the size distribution stems from its influence on the cut-off size (d_{max}), which behaves in accord with Equation (2.112). However, shear also alters the balance between the processes of growth and rupture – the entire distribution adjusting to a new state. Distributions can be described in two ways: either in terms of a distribution function which describes the shape *e.g.* Equation (2.151), or by distribution moments which represent statistical features of the distribution. Examples of moments include the arithmetic mean size (d_L) or the volume average diameter (VAD), representing the average volume of a floc. Sometimes percentiles are employed *e.g.* the median size (d_{L50}), representing the 50th percentile *i.e.* 50% of flocs having size below d_{L50}. The term "d_{max}" noted above is ambiguous in statistical terms, and is often represented by either d_{L95} or d_{V95}, *i.e.* the 95th percentile in the length-size distribution and volume-size distribution respectively. When G is altered, each of these parameters will alter in response to changes in the size distribution.

In some circumstances, one finds that the ratio of the moments *e.g.* VAD/d_L or d_{V95}/d_L remain constant when G is altered. This occurs when all sizes respond to G in the same way *e.g.* all sizes might be halved. If this property exists, the distributions to which it applies are termed as self-similar or self-preserving *i.e.* retaining the same underlying shape. When this occurs, it is much easier to predict the behaviour of all sizes within a distribution. For example, all will behave in the same way as "d_{max}" in response to changes in G. Examples are provided below to illustrate different facets of the response of distributions to changing conditions of shear.

Self-similarity

Figure 3.22 shows co-plots of scaled distributions of alumino-humic flocs subjected to shear. Parameter $u = d/d_L$ is the scaled floc size (d_L defined above) and $\Psi(u)$ is the associated distribution function defined by Equation (2.145). The consistency of shape suggests that the scaled distributions are approximately similar. However, a more sensitive test (in statistical terms) is provided by the behaviour of the moments.

Data in Table 3.11 corresponds to trends shown in Figure 3.22. Judged by the d_{V95}/d_L ratio, the response to G can be roughly separated into two zones shown by the dotted line at $G \sim 60\ s^{-1}$. In the domain $G < 67\ s^{-1}$ there was considerable variation in the moment ratios as the shear conditions were altered, this being particularly evident at low G. If distributions are similar, then the moment ratios must remain constant as G is altered; thus the distributions are not self-similar in this domain. When $G \geq 67\ s^{-1}$, the size ratios are broadly consistent, suggesting self-similarity. In this domain the co-plots shown in Figure (3.22) were fitted by the gamma distribution (Equation 2.151) using the values a = 11, c = 0.25. With this information one may calculate the distribution moments for the fitted distribution using the framework described in Section 2.5.3. These values are shown at the foot of Table 3.11 and are seen to be consistent with the values based on the raw data in the domain $G > 67\ s^{-1}$.

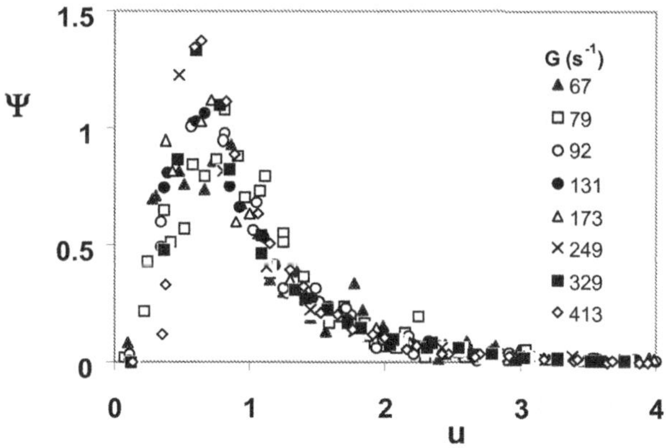

Figure 3.22. Co-plot of scaled distributions of alumino-humic plots subjected to shear (Bache and Rasool, 2006).

Table 3.11. Composite length scales associated with data shown in Figure 3.22 (Bache and Rasool, 2006).

$G\ (s^{-1})$	d_L (μm)	d_{L50}/d_L	d_{L95}/d_L	VAD/d_L	d_{V95}/d_L
8	172	0.87	9.93	2.89	10.8
14	417	3.27	5.69	2.40	6.10
23	568	1.58	2.84	1.50	2.87
33	709	1.38	2.15	1.29	2.26
44	686	1.18	2.41	1.25	2.88
55	643	1.11	2.72	1.28	2.89
67	457	1.24	2.93	1.40	4.04
79	430	1.14	2.54	1.28	3.12
92	363	1.15	3.03	1.38	4.74
131	263	1.17	3.03	1.39	4.11
173	204	1.22	3.45	1.44	4.70
249	164	1.15	3.32	1.45	5.41
329	134	1.06	3.01	1.35	5.21
413	100	1.07	2.83	1.30	4.13
Mean±SD*		1.15± 0.06	3.02± 0.28	1.37± 0.06	4.43± 0.74
Via fitted ψ(u)		1.19	2.86	1.35	4.80

* Based on domain $G \geq 67\ s^{-1}$ and calculated from raw data using a = 11 and c = 0.25 in Equation (2.151).

Mass balance context

When G is altered, and there are no other gains or losses, the solids mass remains constant whatever the shape of the distribution. In cases where the underlying distribution can be modelled by Equation (2.151) and in which the floc solids can be represented by Equation (2.22), the overall solids mass concentration (M) can be expressed by Equation (2.155). As G alters, the overall number of flocs per unit volume (N) follows the trend shown Figure 3.23. Apart from small changes at low G, N is seen to increase continuously with G as a result of the effects of increasing rupture. The conservation statement expressed by Equation (2.155) indicates that if M remains constant, then the product Nd_L^D must be roughly constant, given that other parameters in Equation (2.155) can also vary. The interdependence of N and d_L is evident in Figure 3.23. As G increases, it is seen that d_L passes through a maximum, this occurring at $G \approx 33\ s^{-1}$. In the domain $G > 33\ s^{-1}$, the decrease in d_L with increasing G is driven by the progressive rupture of the larger sizes (*e.g.* d_{V95}). If one wants to maximise floc sizes, it is clear (for this particular example) that G should be selected as $G \approx 33\ s^{-1}$.

When the fitting function $\psi(u)$ is known, one can estimate other parameters of interest using the framework described in Section 2.5. For example, the distribution of mass associated with the plots shown in Figure 3.22 in the domain $G \geq 65\ s^{-1}$ is shown by Figure 3.24. The significance of the condition $G \geq 65\ s^{-1}$, is that the distributions are self-similar in this domain (see above). This implies that the mass distribution can be represented by single plot, *provided* it is represented in scaled form, where $u=d/d_L$. Thus if d_L is known, the association between fractional mass and size is easily determined.

Floc volume fraction

The floc volume fraction (Φ_F) can be roughly estimated by allowing the suspension to settle for about 30 min in an Imhoff cone (or measuring cylinder when the concentration is large).

Alternatively, if the distribution is known, Φ_F can be calculated from the distribution parameters. The simplest approach is via:

$$\Phi_F = \frac{\pi}{6} N\ VAD^3 \tag{3.17}$$

By replacing N using Equation (2.155) and VAD using Equation (2.159), Equation (3.17) is also expressed by:

$$\Phi_F = M \frac{1}{d_o^3 \rho_s A'} \left(\frac{d_L}{d_o}\right)^{3-D} \frac{\Gamma((a+4)/c)}{\Gamma((a+D+1)/c)} \left[\frac{\Gamma((a+1)/c)}{\Gamma((a+2)/c)}\right]^{3-D} \tag{3.18}$$

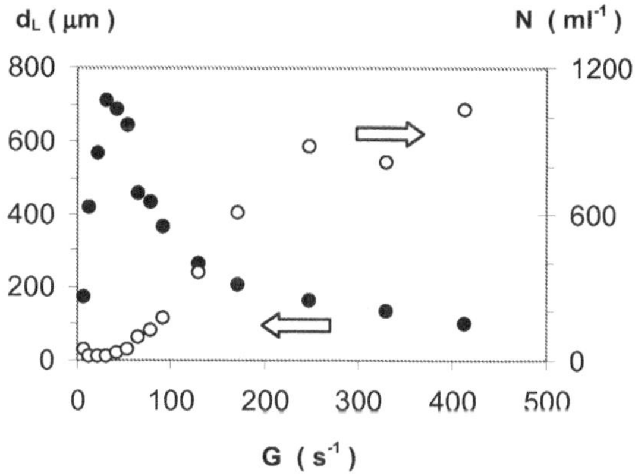

Figure 3.23. Dependence of N and d_L on shear for alumino-humic flocs (Bache and Rasool, 2006).

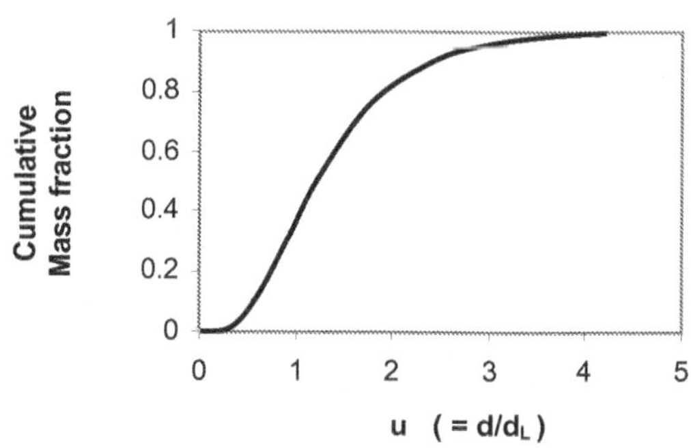

Figure 3.24. Cumulative mass fraction in the shear domain $G \geq 67\ s^{-1}$ (Bache and Rasool, 2006).

Although Equation (3.18) is more complicated than Equation (3.17), it has the merit of providing insight into the parameters that control Φ_F. Of particular note are the dependencies on M and d_L, the remaining features referring to the distribution shape and fractal structure. The plots shown in Figure 3.25 emphasize the sensitivity of Φ_F to shear and the presence of a maximum. Its presence is steered by the existence of a maximum in the d_L-G relationship, such as illustrated in Figure 3.23. Features of this type are likely to influence the potential impacts of G on treatability (*e.g.* Section 3.3). It is seen that the floc volume fraction is sensitive to shear and shows the existence of a maximum.

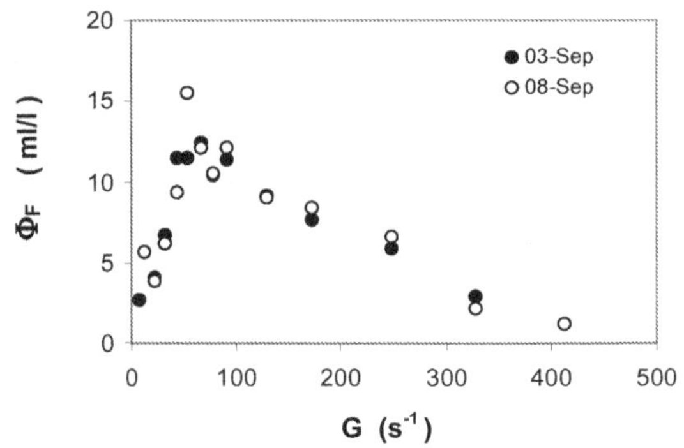

Figure 3.25. Dependence of floc volume concentration on the degree of shear (Bache and Rasool, 2006).

Flocculators at full scale

The data shown Figures 3.22 - 3.25 and Table 3.11 refer to flocs sampled from a water works, and subjected to shear in a Phipps and Bird two-litre beaker of the type illustrated in Figure 3.1. Although there is some evidence of G being a suitable scale-up parameter in geometrically similar systems (de Boer *et al.*, 1989), this is not generally true. The distribution of energy dissipation is likely to be system-dependent and there are many other scaling considerations (see Clark and Feissinger, 1991). In consequence, one cannot assume that optimised G values identified in a small-scale mixer are likely to be applicable in a full size plant. Indeed this is most unlikely (see also Figure 3.27). Nevertheless, there is a reasonable expectation that patterns of behaviour identified in small-scale studies will continue to exist. Ultimately, the impacts of shear are system-specific.

An appraisal of floc size distributions found in a full scale flocculator at a works treating a low turbidity coloured water has been reported in Bache *et al.* (2000) and Bache and Rasool (2006). The coagulant was aluminium sulphate and a polymer floc aid was in use. Details of the flocs were obtained using an underwater camera, backed by image analysis. Shear conditions were altered by changing the rotational speed of the installed paddle system. Values of G cited in Table 3.12 are very rough estimates, but nevertheless indicative of the response to the rotation rate. Distributions were fitted by Equation (2.151), characteristic parameters being shown in Table 3.12. Where there are common values of the fitting coefficients ‘a’ and ‘c’ ($G \leq 48\ s^{-1}$), these indicate similarity in the number-size distributions. There were great difficulties in securing this set because of resolution problems (a warning to others following this path). There is a need to secure far more information of this type, with particular emphasis on the distribution of small sizes.

Table 3.12. Data which describes $\Psi(u)$ in a full scale flocculator (M = 22 mg/l) (Bache and Rasool, 2006).

G (s^{-1})	d_L (μm)	a	c	N (ml^{-1})	Φ_F (ml/l)
13	521	10	0.114	31.6	20.1
25	410	10	0.113	56.4	17.8
48	470	10	0.113	54.2	25.7
102	306	10	0.144	86.0	7.1

3.2.8 Overview

- There is a need to secure a more comprehensive knowledge base of the impact of water quality parameters, coagulants and coagulation practices on floc properties and the ways in which they affect treatment processes.
- Greater emphasis should be given to practices that lead to floc densification.
- Greater attention should be paid to the significance of the packing factor (A′) as distinct from the mass fractal dimension in terms of defining floc structural properties.
- More needs to be learned about the role of coagulant products in the control of floc properties; a particular concern (Section 3.3.7.5) are the polymer-precipitate interactions affecting dewatering.
- There is need to secure more information about floc size distributions at full scale, paying particular attention to the distribution of mass.

3.3 IMPACTS ON TREATMENT PROCESSES

3.3.1 Background

3.3.1.1 Ease of treatment

The performance of treatment processes depends on the nature of the water being treated and the treatment chemistry together with features of design and operational practices. The ease of such solids removal (or treatment) is often referred to as the '*treatability*' of the water. Of the many factors involved, effective coagulation must be singled out as the necessary condition for successful treatment.

Experience indicates that the coagulant dose and pH suiting floc blanket sedimentation also applies to DAF, filtration, sludge thickening and dewatering, when these processes receive a freshly coagulated suspension. However, as coagulated water passes through a plant, the solids stream is affected by the time factors (Section 3.2.7.2), changing levels of shear (Section 3.2.7.3) and changes in chemistry. These affect the nature of the floc and its treatability. Clues concerning treatability can be gleaned from design and performance data, such as presented in Table 3.13. The data show that soft coloured waters behave very differently from hard, turbid waters, but why? One clue lies in the solids content of the thickened sludge - the sludge from soft coloured waters having a higher water content (probably caused by a greater abundance of bulk precipitate). Precipitate abundance is likely to be responsible for the greater polymer dose in the soft coloured water (Section 3.3.7.5).

Table 3.13. Performance benchmark reported values compared with empirical design values for continuous sludge thickening (Dillon, 2004).

Benchmark	**Typical/design value**	**Actual values Average (range)**
Polymer dose, kg/t	1-2 (hard, turbid sources) 2-4 (soft, coloured waters)	1.47 (0.10-5.20) 3.18 (0.07-17.1)
Hydraulic loading rate, $m^3/m^2.h$	1.50 (hard, turbid sources) 0.75-1.50 (soft, coloured sources)	1.58 (0.50-6.53) 0.87 (0.22-1.80)
Solids loading rate, $kg/m^2.h$	2.0-4.0 (soft, coloured sources) 4.0 (hard, turbid sources)	1.7 (0.3-2.9) 2.4 (0.2-4.0)
Thickened sludge solids conc. %w/w	2.5-5 (soft, coloured sources) 5-10 (hard, turbid sources)	2.7 (1.0-7.9) 4.0 (0.5-12.0)
Thickener supernatant turbidity, NTU	4-8 (hard, turbid water sources) 4-8 (soft, coloured sources)	7.1 (1.7-25.0) 13.6 (0.2-50.0)

The sludge water content is tied to the water content of the source floc. A greater water content implies a lower solids concentration and diminished strength (Section 2.3.5). Table 3.13 indicates that the hydraulic loading rate of continuous sludge thickeners is lower in soft, coloured waters than in hard turbid waters – a feature which is presumed to emanate from the character of the source floc. Unconditioned soft coloured waters tend to be characterised by a 'weak floc' of high moisture content. Thus it appears that the hydraulic loading is linked in some way to floc strength. If strength were used as a benchmark for characterising flocs, it would help to account for observed differences in treatability. Treatability is often identified by the hydraulic loading rate (*i.e.* treatment rate) of a process.

A classification scheme which attempts to link treatment rate and strength is suggested in Table 3.14. The content of Table 3.14 (and its supporting notes) is based on the author's (RG) experiences gained from pilot plant studies and at full scale.

Table 3.14. Typical base-line treatment rates.

Process	Floc quality	FBC (m/h)	DAF (m/h)	Filtration (m/h)
Low	Weak	1-2	5-8	4-6
Medium	Moderate	2-4	8-12	5-8
High	Strong	3-6	10-14	6-10

Notes:

(i) Lowest and highest treatment rates depend on water temperature, though DAF is perceived as being relatively insensitive to water temperature.

(ii) Appropriate choice and application of a polyelectrolyte can increase treatment rates by about 30% or more.

(iii) Some proprietary DAF units are referred to as "high-rate" and claimed to operate at rates up to about twice the above. Their performance will be more sensitive to control of flocculation and effect of water temperature.

(iv) Higher filtration rates, up to about twice the above, are possible when good clarified water quality can be assured.

3.3.1.2 Processes as collectors

At first sight flocculators, FBCs, DAF and filters appear to be very different types of process. However [as encountered at water works] they have much in common, in that in each case the incoming stream of flocs encounters an array of collectors. In the case of flocculators and FBCs, the collectors are a quasi stationary distribution of flocs, the collection process coming under the heading of contact flocculation as described in Tambo and Hozumi (1979). In the case of DAF the collectors are bubbles, and for filters, the collector is the packed bed. The rate of loss of particles from the incoming stream depends on the flux of particles past a collector (involving the collision cross-section), the attachment efficiency (η_{pc}) and the particle-collector collision efficiency (α_{pc}). Schematically, the loss of concentration of incoming particles in passing through an 'absorbing layer' of collectors can be represented by:

$$\frac{dn_p}{dt} = -\alpha_{pc}\eta_{pc} k\, n_p n_c \tag{3.19}$$

in which n_p is the number of incoming particles/flocs per unit volume, n_c the number of collectors per unit volume and k a transport coefficient with units ($m^3\,s^{-1}$). Examples of Equation (3.19) are provided by Equations (1.11) and (1.19) with $\alpha \equiv \alpha_{pc}\,\eta_{pc}$. The mass transfer rate constant depends on the prevailing transport mechanism *e.g.* diffusion, turbulent transfer, sedimentation or local stream flow. In Equation (1.12), k has no size dependence, whereas in Equation (1.18) $k \propto d^3$. When particles/flocs are larger than 1 μm (*i.e.* above the regime controlled by molecular diffusion), k increases with size - the precise dependence depending on the transfer mechanism. The collision efficiency depends on both relative sizes of the particles involved, but tends to follow the trend shown in Figure 3.26.

The important features of this trend is the appearance of the minimum at about 1 μm and the increase with particles size when d > 1 μm. The attachment efficiency or 'stickability' depends on colloidal forces, such as indicated in Section 1.3.3.2.

From this rather compressed review, the following points emerge:

- removal is most difficult for particles of about 1 μm
- when d > 1 μm , the removal efficiency (represented by the combined term $\alpha_{pc}\eta_{pc}kn_c$) increases with particle size, concentration of collectors and the attachment efficiency.
- Treatability is generally enhanced by practices that lead to increases in size, strength and attachment efficiency.

Attachment is favoured by effective coagulation. By itself, increasing size is not necessarily beneficial, because of reductions in strength. Equally, small flocs may be stronger, but can lack the necessary size for efficient removal. What is often needed is the combination of density, strength and size. This is favoured by optimal coagulation, good dosing practices, floc aids and 'tuned' flocculation. The sizes of greatest impact are likely to be within the range 1 to 100 μm.

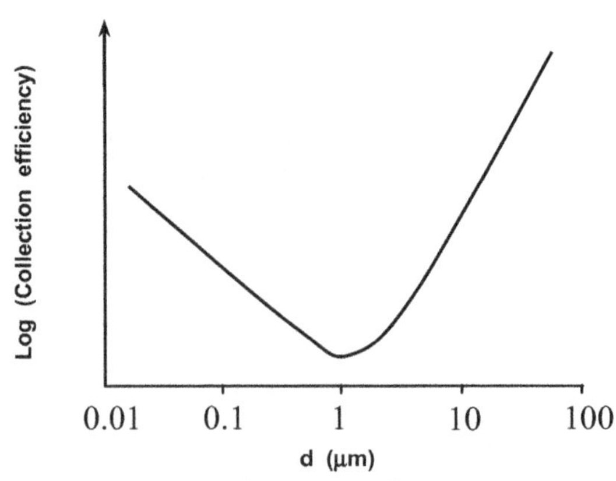

Figure 3.26. Schematic dependence of collision efficiency as a function of particle size for a spherical collecting particle (based on Probstein, 1994).

3.3.1.3 Thrust of Section 3.3

Beyond the introductory remarks given above, knowledge of flocs in treatment processes is somewhat lacking. Much has been written about the design and operation of individual processes, and each has been subjected to considerable scrutiny. Here the task is not to reproduce such information, but rather to illustrate (where we can) how floc and material properties affect treatment processes. The account is scanty and fragmentary. In places, the central theme is broadened to include related issues affecting practice.

3.3.2 Chemical application

The manner in which chemicals are added affects coagulation, and in consequence is likely to affect floc development. Some guidance is provided in Table 3.15, but this should not be interpreted too literally as there are often compounding factors.

Order of dosing

There has been much debate about the order of chemical addition and pH-adjusting chemicals. Some support coagulation addition before base addition (Nickel, 1945; Kawamura, 1973), whereas others support the reverse (Moffett, 1968; Jeffcoat and Singley, 1975).

When using hydrolysing coagulants (particularly alum), the most reasonable chemical application sequence is to first lower the pH of the raw water through the addition of alum (or even with acid) to form more highly charged species for reaction with colloids. The next step is to adjust the pH of the water to the range for minimum aluminium or iron solubility to facilitate floc formation. PAC (powdered activated carbon) and potassium permanganate are more effective when fed upstream of the alum. Chlorine can be fed either upstream or at the same point as alum. If pH and alkali deficiency need to be corrected, lime and caustic soda can be more effective if they are fed at least several seconds after the alum feed point; this applies to lime water or high quality 'milk of lime'. However, for poor quality lime (with larger quantity of inert solids present), or for direct filtration (when the time available or chemical dispersion and micro-floc development is restricted), it may be better to dose before, or at same time as the coagulant.

A particularly good illustration of the benefit of post coagulation lime addition is described in Gregor, Nokes and Fenton (1997) who were concerned with the removal of NOM from low turbidity waters by alum coagulation. The problem with such waters is that they can lack sufficient nucleation sites for floc development. The authors found it advantageous to adjust the pH downwards to between 4 and 5 prior to coagulant addition, this encouraging the formation of soluble NOM-aluminium complexes.

Table 3.15. Issues affecting dosing practices.

Factor	References
Order of dosing	
There appears to be wide agreement arising concerning order of dosing of the chemicals required for coagulation, *i.e.* acid-coagulant-base. However, the benefit might not always be evident. If this is the case, it may indicate poor dispersion, or that the dose and/or pH are not optimal. • Acid (whether for optimising coagulation or other reasons) should be dosed before coagulant. • Coagulant should be dosed before base. However in some situations, dosing calcium before coagulant may be useful. Also, dosing alkali first might be better in a fast system such as direct fitration. • Dose PAC or permanganate before coagulant. • Dose cationic polymer coagulant at same time as metal-ion coagulant, otherwise simultaneous dosing of chemicals should be avoided.	Section 3.3.2
Rapid mixing	
Vital for dispersion; can lower required dose when destabilising by charge neutralisation	Section 3.2.7.1
Time lag/ delay time	
Coagulants: The delay time between chemical dosing and inlet to a floc blanket (to allow flocculation in connecting channels) affects the efficiency of an FBC.	Yadav and West (1975)
Polymer flocculants: Time lags between dosing of coagulant & flocculant and between flocculant & inlet of FBC (to allow flocculation in connecting channels) improves removal efficiency.	Table 3.10
Concentration of chemical	
Dilution of coagulant prior to dosing enables more accurate dosing and benefits chemical dispersion – providing dilution is not great enough to cause premature hydrolysis	Dempsey (1984) Gregory *et al.* (1996a)
Alkali	
Choice of alkali depends on costs (capital + operating). Lime may be preferable to caustic soda when treating low turbidity waters, even if more costly.	Section 3.3.2

The optimum pH for soluble NOM removal and minimum aluminium residuals was between 6.0 and 7.0 for the waters under study. As the pH is raised, the soluble complexes link with each other to form larger insoluble complexes (micro-flocs) that serve as nuclei for macro-floc development. When lime was added immediately after coagulation (to fulfil the necessary adjustments to pH and alkalinity), it was found that floc settlement improved. In contrast, when caustic soda was used for pH control, this did not affect settlement. The difference in behaviour was attributed to the particulate nature of lime, this accelerating floc formation (due to greater particle concentrations) and settlement (probably due to larger flocs and increased solids density).

In contrast to the above, the addition of calcium ions before a coagulant, can ameliorate flocculation problems associated with extracellular organic matter (EOM) formed by algae. Calcium ions neutralise the negative ions of the polymer anions and become bound as calcium-EOM. This results in a reduction of the potential of the EOM to form complexes with metal-hydroxo complexes. The resultant aggregates thus formed are easier to filter after flocculation (Bernhardt *et al.*, 1986). Using a pressure filter pilot plant treating a soft and coloured upland water, the author (RG) found that longer filter runs were obtained by dosing the alkali (sodium carbonate) first. Run length appeared to be affected not only by the order of addition (and possibly by the delay time between the addition of alum and alkali), but also by the position at which the second chemical was added, this determining the point at which flocculation could commence.

Chemicals which should not be simultaneously fed into a flash mixer include alum and anionic or non-ionic polymers, alum and lime or caustic soda, chlorine and PAC, and alum and PAC (Kawamura, 1996). Ammary and Cleasby (1997) note that dual coagulants can have an advantage over the use of an inorganic coagulant or a cationic polymer when used alone. When using this mix, the chemicals should be added simultaneously.

3.3.3 Flocculation

In a standard treatment sequence in which the system chemistry has been optimised and necessary aids applied, the task of flocculation is to render the solids phase into a state which enables effective separation. Depending on the nature of the raw water and the scale of the plant, commonly used separation systems downstream of flocculators include floc blanket clarifier (FBC), dissolved air flotation (DAF) and direct filtration units. Clarifiers are usually backed by sand filters - the latter also being sensitive to flocculation practices in spite of being further along the process chain.

Concepts for describing flocculation are outlined in Section 1.3.3. Traditional design practice has tended to focus on control of the velocity gradient, G, and the retention time, τ, these manifesting in the form of the product $G\tau$ known as the Camp Number. G is often set in the range 20-70 s^{-1} and τ in the range 10-40 minutes. According to Kawamura (1996), experience indicates that the choice of the most appropriate $G\tau$ value is *site specific*, although many books cite a range of 10^4 to 10^5, this representing a condition in which settled water turbidity is minimised. This immediately raises the question as to whether a criterion tied to sedimentation behaviour is appropriate for DAF or direct filtration. Before probing this issue, it is worth examining the significance of the component terms and their impact on floc development.

With regard to the retention time, there is merit in the suggestion of Lawler and Nason (2004) that the design of flocculators should be based on the characteristic reaction time for the loss of 1 μm particles. Thereafter the retention time can be based on some multiplier of this characteristic reaction time. The multiplier would reflect flow patterns within a flocculator (affecting the effective retention time) and the goals of the flocculation stage. The focus on sizes ~ 1 μm is appropriate because this is representative of the size range which is most difficult to remove in *any* separation process by virtue of its low collection efficiency and low sedimentation velocity. If the retention time is too small, one will fail to integrate the microflocs into the macrofloc and clearly there is little to be gained by extending retention times beyond what is necessary to fulfil this task.

The velocity gradient couples with the retention time in terms of controlling the growth process, or conversely, controlling the loss (or collection) of microfloc as noted above. Alterations in G also offer the opportunity of controlling features of the floc size distribution and structure. Changes in structure are more difficult to control, tapered flocculation offering opportunities (Section 3.2.7.2). In its traditional role, the velocity gradient is used as a scale-up parameter through its control on the volume average value of the energy dissipation per unit mass. While its potential as a scale-up parameter has been examined and supported in studies such as de Boer *et al.* (1989), it must be recognised that it is one of many parameters that can be adopted for scale-up purposes (see Clark and Fiessinger, 1991).

The problems of scale-up are illustrated in Figure 3.27. This shows a comparison of the maximum size of sampled flocs placed in a laboratory scale mixer with that found in a full scale flocculator, the comparison applying to a series of water works treating low turbidity coloured water. In each case the value of G in the laboratory mixer matched that found at full scale (insofar as G could be estimated). It is seen that the sizes compare favourably when the floc is weak, but show marked differences when a floc aid was present. It should be recognised that scaling by G does not deliver corresponding time scales - the opportunities for growth being greater in a larger reactor with longer circulation times. Further, scaling by G does not exercise control on the *distribution* of energy dissipation within a mixer - this depending on the system design. For corresponding G values, differing regimes of energy dissipation are likely to cause differences in both mean size and structure. The effect of polymer illustrated in Figure 3.27 may reflect the time at which the polymer dose was applied relative to development of floc mean size, amongst other possible reasons. It should also be emphasised that flocculation is not confined to the flocculators or to rapid mix units, the process taking place at all times following coagulant addition. An audit of the energy loss (and corresponding G values) prior to the clarifiers has been reported in Gregory (1987).

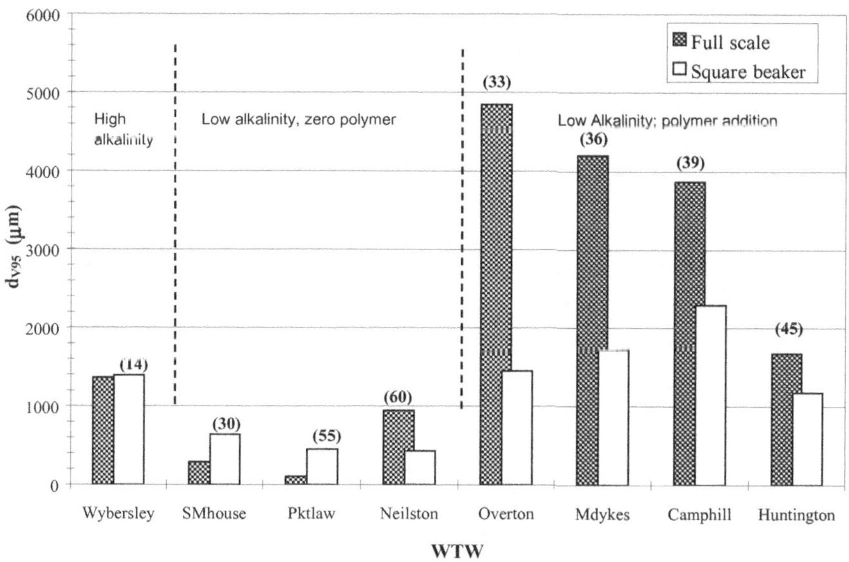

Figure 3.27. Comparison of d_{V95} full scale with square beaker data at comparable G values (shown in brackets) (Bache *et al.*, 2000).

The upshot of these remarks is that there is no guarantee that optimised behaviour can be delivered by blindly following criteria such as the Camp number or similar formulae cited in the literature. In this respect, the observation of Karamura (1996) noted above is entirely appropriate. Such criteria should be used as guidelines as possible starting points for optimising.

We now come to the vexed question of what should be delivered by flocculation. It is a question that can only be answered in general terms because it depends on the nature of the water and the separation processes that are in place.

In a discussion of solid-liquid separation process selection, O' Melia (1985) indicates that particles > 100 μm will be completely removed by settling (based on a particular scenario). Although a criterion of this type represents a useful target, it is not always obvious how a selected size affects separation efficiency in the context of the entire size distribution. For example, should a 100 μm size be assigned to d_L, (the arithmetic average size), or to some other representative size within the distribution?

Filter beds are least efficient when they are clean. At this condition, O'Melia (1985) suggests that particles > 10 μm can be effectively removed. Filter straining becomes significant when the ratio of the particle size to the filter media size > 0.2. Therefore, assuming an average media size of 0.5 mm for a sand filter, flocs > 100 μm should not be applied to this type of filter. In direct filtration, with coarser media, larger flocs can be accommodated. Beside size, consideration must be given to the volume of material being delivered to a filter, because it is essentially the trapped volume that drives the head loss (Section 3.3.6). Basically one should be seeking to minimize the floc volume fraction (*e.g.* Figure 3.25), and at the same time endeavour to deliver sizes in accord with the guidelines noted above.

Concerning FBCs, Yadav and West (1975) found that the magnitude of a delay time (between chemical addition and entry into an FBC) had a significant impact on removal efficiency of a floc blanket; the degree of agitation during the delay time was relatively unimportant. Reasons were not established, but must be tied to floc development in the delivery channels; see also the discussion of Figure 3.31 in Section 3.3.6.4.

DAF units differ greatly from FBCs in terms of the stresses faced by flocs. On the basis of both modelling and experimental investigation, Edzwald (1995) recommends flocculators used in drinking water applications of DAF should be designed to produce strong floc with particle size distributions in the range 10-30 μm. If one accepts this criterion, it is unnecessary to produce larger floc, only to have them ruptured in regions of high energy dissipation within DAF units. Valade *et al.* (1996) suggests that commonly used flocculation times can be considerably reduced (as low as 5 min) because experience indicates that coagulation chemistry (pH and coagulation dose) has a stronger influence on DAF performance than physical flocculation parameters (notwithstanding rapid dispersion). Thus one might speculate that the task could be achieved by using well-controlled rapid mixing coupled with growth in distribution channels. However, low water temperature may be an issue.

In all separation processes, it is vital that flocs can withstand the hydrodynamic stresses they encounter. In general, this attribute is favoured by using smaller sizes, these offering a higher solids volume concentration and greater strength.

3.3.4 Floc blanket clarifiers

The dynamics of floc blanket clarifiers have been described in studies/reviews such as Ives (1968), Gregory (1977) and Gregory *et al.* (1996b). The floc blanket is best described as a fluidised bed.

When the bed height (the top of the blanket) is stationary, there is a balance between the net gravitational forces acting on the blanket and the drag forces exerted on the flocs by the upward flow. As micro-flocs pass through the blanket they collide with the macro-floc - capture being controlled by the rate of collision and the trapping efficiency. When there are no losses, solids accumulation within the bed generally manifest as an increase in the bed height. Thus if the bed height is stationary, it also reflects a balance between the solids input and output, the latter being controlled by sludge withdrawal from the bed, but some escaping as 'carry-over' in the effluent stream. Generally the objective is to maximise the liquid throughput, this being constrained by the hydrodynamic stability of the bed, and the desired separation efficiency of the process. In the foregoing, we provide a simplified description of kinematic features of an FBC in order to highlight the controlling parameters and to identify potential impacts of floc properties.

In any layer of mass concentration (c), blanket stability is controlled by a balance between the upward solids flux (uc) and the downward settling flux ($v_{sett}c$). The schematic dependence of these components on c is illustrated in Figure 3.28.

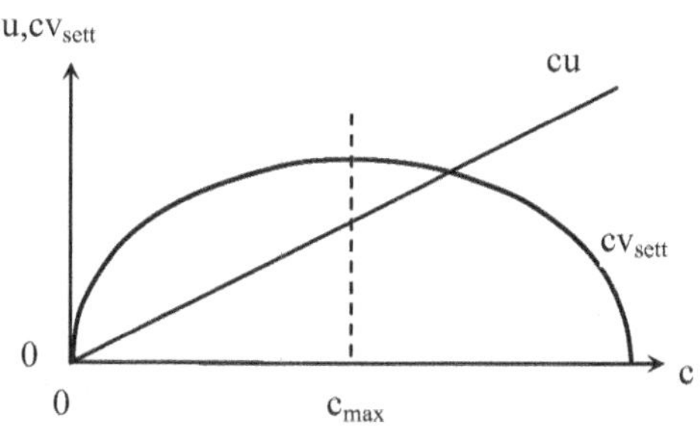

Figure 3.28. Schematic dependence of solids flux components cu and cv_{sett} on solids concentration.

In the case of the upward solids flux, there is a linear dependence with u defined by the superficial velocity (Q/A). In the case of the settling flux, its value tends to zero at low concentrations because $c \rightarrow 0$ and at high concentration because $v_{sett} \rightarrow 0$. Between these extremes, $v_{sett}c$ passes through a maximum at c_{max}. Ideally, if one can operate at c_{max}, there is a prospect of maximising the solids throughput (Gregory *et al.*, 1996b). To simplify discussion, we treat the bed as quasi-uniform, the change in the position of the blanket surface being determined by:

$$\frac{dL}{dt} = u - v_{sett} \tag{3.20}$$

with L referring to the height of the blanket above the bottom of the tank (*e.g.* Head *et al.*, 1997). In the stationary state, $u = v_{sett}$; this shows that increases in u can only be sustained by increasing v_{sett} within the domain $c \leq c_{max}$.

Experience shows that the blanket floc volume fraction is likely to be within in the range 10 – 30 %, implying that one is dealing with hindered settlement as distinct from discrete particle settlement. In hindered settlement, v_{sett} can be represented by an empirical relationship of the form:

$$v_{sett} = v_o(1 - k\Phi)^p \tag{3.21}$$

in which v_o is a scaling velocity and k, p (both > 0) are fitting coefficients. In Equation (3.21), the term Φ is normally associated with the solids volume fraction (ϕ), but here refers to the floc volume fraction (Φ_F). Where suspensions are uniform, v_o can be specified by the free settling velocity of an individual particle. In the case of a heterogeneous flocculant suspension, v_o cannot be readily identified in physical terms. However, in a plot of v_{sett} versus Φ, it can be estimated by the projected terminal settling velocity at the condition $\Phi \rightarrow 0$. Given the loading rates used in practice (Table 3.14; implying low settling speeds), v_o is likely to comply with the form of Equation (2.67) *i.e.* $v_o \propto \mu^{-1}$ in which μ is the dynamic viscosity (Gregory, 1977); the viscosity impact is of concern when one considers the impact of temperature on settlement and thereby on the surface loading.

Two features of a fluidised bed warrant attention. First, when u increases, the bed expands. If we assume that the floc volume is conserved (not necessarily accurate, but satisfactory as a first approximation), it implies that $V\Phi_F$ is constant, where V is the bed volume (=LA). Hence one may write when cross-sectional area A is constant:

$$\frac{\Phi_F}{\Phi_{Fo}} = \frac{L_o}{L} \tag{3.22}$$

in which the subscript 'o' refers to an arbitrary reference condition. In a tank of regular shape *e.g.* pyramidal rather than parallel-sided, A is a function of L and therefore Φ_F remains proportional to L. The main point to observe is that the floc volume fraction alters when the bed depth changes. Second, the pressure drop (ΔP) across a fluidised bed is equal to the buoyant weight of the floc *i.e.*

$$\Delta P = h\rho_w g = L\Phi_F < \rho_e > g \tag{3.23}$$

in which h is the head loss and $<\rho_e>$ the average density difference (ρ_f - ρ_w) within the blanket. Because $L\Phi_F$ is assumed to be conserved, it is seen that the pressure loss is independent of the bed expansion (*i.e.* increases in bed depth), and thereby of the upflow velocity.

Drawing on earlier work of Ives (1968) and Tambo and Hozumi (1979), Amirtharajah and O'Melia (1990) considered that when the blanket behaved as a turbulent reactor, the effluent concentration of micro-flocs could be estimated using the expression:

$$\frac{n}{n_o} = \exp\left(-\frac{\alpha G \Phi_F t_L}{\pi}\right) \tag{3.24}$$

in which n_o is the number concentration of micro-flocs entering the base of the blanket, α the trapping efficiency and t_L the liquid retention time. The trapping efficiency depends on factors noted in Sections 1.3.3.3 and 3.3.1.2. Equation (3.24) shows that the effluent concentration is reduced by increasing the value of $\alpha G\Phi_F t_L$, in which the liquid retention time is defined by:

$$t_L = \frac{(1-\Phi_F)V}{Q} = \frac{(1-\Phi_F)L}{u} \tag{3.25}$$

From knowledge of the pressure drop across the blanket, G is estimated by:

$$G = \left(\frac{\Delta P}{\mu t_L}\right)^{1/2} = \left(\frac{u\Phi_F g < \rho_e >}{(1-\Phi_F)\nu\rho_w}\right)^{1/2} \tag{3.26}$$

where ν is the kinematic viscosity of water. With this information, the attenuation term $G\Phi_F t_L$ appearing in Equation (3.24) can be evaluated. If we recognise that $u = v_{sett}$, with v_{sett} defined by Equation (3.21), and assume that $L \propto u$ in the fluidised state (see Figure 3 in Mutl *et al.*, 2002), then attenuation is steered by the dependence:

$$G\Phi_F t_L \propto \Phi_F^{3/2}(1-\Phi_F)^{1/2}(1-k\Phi_F)^{p/2} \tag{3.27}$$

To illustrate the implications of this relationship, data cited in Figure 6 of Gregory *et al.* (1996b) were fitted by Equation (3.21) to yield the values v_o = 3.5 m h^{-1}, k = 2.9, p = 1.16, the maximum settling flux occurring at $\Phi_F = 0.16$. In the range $0 < \Phi_F < 0.16$, the model indicates that $G\Phi_F t_L$ increases continuously with increasing values of Φ_F, implying that the separation efficiency increases with the floc volume fraction. It should also be noted that $G\Phi_F t_L \rightarrow 0$ as $k\Phi_F \rightarrow 1$, indicating that $G\Phi_F t_L$ passes through a maximum. Whether this mathematical feature has any physical significance is open to question. The assumption $L \propto u$ noted above (in conjunction with Equation 3.22) implies $\Phi_F \propto u^{-1}$ (roughly consistent with trends reported in Gregory, 1979). A reduction of Φ_F with increasing u, means that the product $G\Phi_F t_L$ diminishes. On the basis of Equation (3.24) (with α remaining constant), this trend suggests that the quality of the water leaving the blanket will deteriorate as the upflow velocity is increased - a feature which is observed in practice.

The mechanics of removal and dynamical features of the blanket are far more complicated than described above (see for example Su *et al.*, 2004). The model applies to a steady state regime in which the solids mass within the blanket remains constant (a consequence of dL/dt = 0 in Equation 3.20). It neglects the interaction between the floc volume concentration and shear (*e.g.* Figure 3.25). There is no direct provision for the effects of floc size on either the trapping efficiency or a form of stratification also known as *segregation* examined in Pruden and Epstein (1964). In the case of FBCs, Gregory *et al.* (1999) note that segregation occurs at low upflow rates, with faster settling denser floc accumulating in the lower part of a blanket and the slower settling floc in the upper part of the blanket. When the upflow rate is great enough, segregation disappears. The settled water quality can reflect the presence and absence of segregation, and in particular it reflects the concentration of the solids in the upper blanket.

The role and impact of the blanket floc volume fraction is emphasised in Gregory *et al.* (1996b) and must be regarded as central to the dynamics of the FBC. A practical problem is to determine Φ_F. Based on much experience, the approach advocated in Gregory *et al.* (1996b), is to identify Φ_F as the fraction of a sample volume occupied by the settled floc after 30 min settlement *in a 100 ml measuring cylinder.* At maximum flux, Φ_F has been found to be in the range 16 to 20% for alum coagulation and tending to be greater (possibly 25 to 30%) when polyelectrolyte is used as a flocculant aid or when using an iron coagulant. The actual value depends on the water quality and the coagulation chemistry.

As indicated above, an effect of a polyelectrolyte is to increase the blanket solids concentration, with the prospect of increasing the attenuation efficiency. However Gregory (1977) found that the attenuation efficiency approached a limiting value as the polyelectrolyte dose was increased. Commenting on this feature, Gregory *et al.* (1999) report that there is little point in applying polyelectrolyte (or increasing the dose) if the blanket floc concentration is greater than about 18%. The reason for this limiting condition has not been established, but is presumed to be some manifestation of floc character, such as the mechanism of micro-floc removal in a blanket or may be tied to the polymer adsorption capacity of the floc (Section 3.3.7.5).

From the review it is seen that the principal factor controlling the surface loading is the magnitude of v_{sett}. Its magnitude (and thereby the surface loading) can be increased by use of flocculation aids. In practice, the blanket becomes unstable when the upflow velocity exceeds about $0.7v_o$ (Gregory, 1977; Gregory *et al.* 1999). The dependence $v_o \propto \mu^{-1}$, implies that the surface loading will be reduced in cold conditions as a result of increased viscosity (Table 3.6). The removal efficiency increases with Φ_F, whose optimum value is defined by the maximum flux condition; its maximum value relates to the gel point and the onset of compression and thickening. As well as affecting settlement, the incoming floc size affects the collection efficiency. Some practical aspects of blanket control are summarised in Table 3.16. Further detail is provided in Gregory *et al.* (1999).

Table 3.16. Examples of practical issues in blanket control.

Symptom	Causes/Remedies
'Boiling' (in some FBCs)	Inequality of flow distribution or imbalance of flow geometry
Blanket level	
Distinct interface	Should be able to see/detect distinct surface of blanket
Diffuse interface	May indicate that "maximum flux condition" is being exceeded
High blanket level	Need to draw off solids
Low blanket level	Not enough retention in blanket (*i.e.* low $G\Phi_F t_L$ value): Suspend withdrawal of solids
Blanket not formed (carryover concentration similar to influent)	Flow rate may be too high; try reducing rate Import floc from adjacent clarifier Review chemistry and dosage Temporary use of ballasting agent
High concentration in carryover (>influent concentration)	May mean blanket level control inadequate (*e.g.* capacity of sludge cone not great enough) Can result from rate being increased too quickly

3.3.5 Dissolved air flotation

From origins in the mineral processing industry, DAF is now used widely within WTWs as a means of clarification. Whereas FBCs are limited by settling velocities (with upflow velocity normally below 6 m h^{-1}, DAF units are now being used at flow rates of 20 to 30 m h^{-1} and with the prospect of even higher values (Kiuru, 2001). As a result of high throughput they offer a smaller footprint than FBCs and offer a greater degree of operational control. Beside economic advantages, DAF units will often represent the best choice for waters producing flocs of low settling velocities, examples including eutrophic algal-rich waters and soft, low alkalinity upland waters.

The separation process can be considered as having two stages. The first stage takes place within the contact or reaction zone (see Figure 3.29). In this phase, the incoming stream of flocs encounters the bubble stream, which acts as collectors of particles. The efficiency of this process is vital in terms of its impact on quality of the clarified water. With bubbles attached, flocs are conveyed to the upper surface as a result of their buoyancy, the clarified water being drawn off at the base of the tank in the *separation zone*. The dynamics of this second stage are the inverse of a sedimentation tank, with the 'downflow' velocity (*i.e.* the surface loading) controlling the degree of separation.

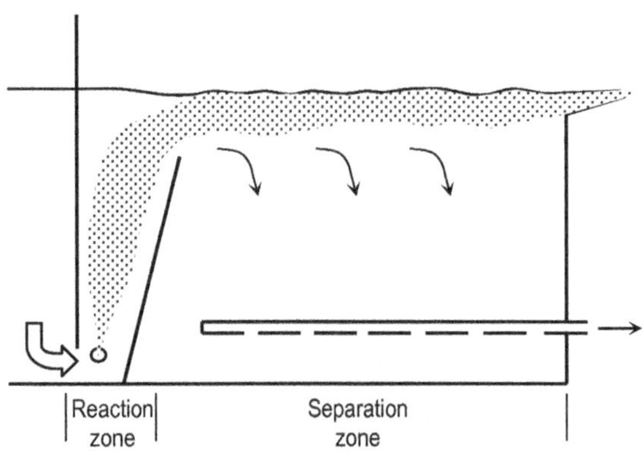

Figure 3.29. Conventional DAF unit (5 - 7 m h^{-1}).

The aim is to generate large bubble number concentrations (~ 10^5 ml^{-1}) of small bubbles by injecting supersaturated water into the base of the contact zone. In theory, the larger the saturation pressure, the smaller the critical diameter of the bubble nuclei (Edzwald, 1995). However, many other factors, such as nozzle design and the mechanism of bubble nucleation affect the bubble size. In practice, saturation pressures of 400 – 600 kPa (4 – 6 bar) lead to the formation of bubbles in the size range 10 to 100 μm with 40 μm serving as a representative value (Edzwald, 1995).

Bubbles can become attached to particles in various ways, particle collisions and adhesion with preformed bubbles being the major mechanisms of concern. Successful adhesion requires that interacting surfaces are conducive to bond formation. Edzwald suggests that this is favoured by neutrally charged particles and particle surfaces, which are partially or wholly hydrophobic. A quite different view is given in Fukushi *et al.* (1995) and Han (2002), whose measurements show that emerging bubbles are negatively charged. On this basis they bond most readily on positively charged sites. Though freshly formed precipitates from metal coagulants have this characteristic, Edzwald (1995) comments that the presence of a bound water layer around hydrophilic particles hinders bubble adhesion. Beyond this, bubbles and particles interact in the same way as any heterogeneous group of suspended particles *i.e.* they are subject to hydrodynamic and colloidal forces. Given that the smallest size of incoming flocculated solids is likely to be a few μm, Edzwald indicates that interception is the dominant contact mechanism.

There are two basic approaches for modelling the collection process. In the first *e.g.* Edzwald (1995), laminar flow conditions are assumed. A more complex model (based on orthokinetic flocculation) is described in Fukushi *et al.* (1995). Of the two, the latter is more realistic, but difficult to calibrate. In the much simpler Edzwald scheme (based on Equation 3.19), the component factors controlling the trapping efficiency of particles by a single bubble are expressed by $\alpha_{bp}\eta_{bp}$ in which α_{bp} is the particle-bubble adhesion efficiency and η_{bp} the trapping efficiency of bubble as a single collector (affected by both electrical and hydrodynamic interactions). If we envisage an array of bubbles rising with velocity v_b relative to the surroundings, the total volume swept out per unit time by the integrated bubble cross-section is $n_bA_bv_b$ in which A_b is the average cross-sectional area of a single bubble and n_b the bubble number concentration per unit volume. Within this volume, the number concentration of particles (or flocs) per unit volume is n_p. With $\alpha_{bp}\eta_{bp}$ as the overall collection efficiency, the number of particles trapped per unit time is given by:

$$\frac{dn_p}{dt} = -\alpha_{bp}\eta_{bp}n_bA_b\mathrm{v}_bn_p \tag{3.28}$$

By introducing the transformations $A_b = \pi d_b^2/4$ with d_b as the bubble diameter, $n_b = \phi_b/(\pi d_b^3/6)$ with ϕ_b as the bubble volume fraction (vol/vol) and $dn_p/dt = v_b\ dn_b/dz$, Equation (3.28) becomes:

$$\frac{dn_p}{dz} = -\frac{3\alpha_{bp}\eta_{bp}\phi_b}{2d_b} n_p \tag{3.29}$$

Integration of Equation (3.29) shows that the collection efficiency (E_c) across a contact zone of length L_c is expressed by:

$$E_c = 1 - \exp\left(-\frac{3\alpha_{bp}\eta_{bp}\phi_b L_c}{2d_b}\right) \tag{3.30}$$

This provides insight into the critical parameters affecting the trapping process. Water quality factors and chemical conditioning affects α_{bp}. Aside from colloidal force interactions, η_{bp} tends to be maximised when bubbles and particles have similar size (see Han, 2002). Bubble sizes tend to be fixed, as indeed L_c showing that the bubble volume fraction can function as a control parameter.

After attachment, the impact of bubbles on the flotation of a particle cluster can be obtained from a balance of the forces acting on a particle-bubble combination. In the case of a single particle within the Stokes regime (Re < 0.5), the force balance may be stated as:

$$3\pi\mu d_{pb} \mathrm{v}_{\mathrm{pb}} = \frac{\pi}{6} g\left[n_{pb} d_b^3 \rho_w - d_p^3(\rho_p - \rho_w)\right] \tag{3.31}$$

in which v_{pb} is the rising velocity of the particle-bubble combination, n_{pb} the number of bubbles attached to the particle and d_{pb} the effective size of the particle-bubble combination. The latter is defined by:

$$d_{pb} \approx (n_{pb} d_b^3 + d_p^3)^{1/3} \tag{3.32}$$

When the 'particle' is a floc, account must be taken of its fractal structure. Making use of Equation (2.24), v_{pb} is specified by:

$$\mathrm{v}_{pb} \approx \frac{1}{18}\frac{g d_{pb}^2 \Delta\rho}{\mu} \quad \text{with} \quad \Delta\rho = n_{pb}\rho_w \left(\frac{d_b}{d_{pb}}\right)^3 - A'(\rho_s - \rho_w)\left(\frac{d_o^3}{d_{pb}^3}\right)\left(\frac{d_p}{d_o}\right)^D$$

...(3.33), (3.34)

in which ρ_s is the solids density within the floc, subscript 'p' now referring to the floc; this is only valid in the Stokes range. Outside the Stokes range, computations must be based on a force balance in the form of Equation (2.62).

By replacing the drag coefficient (C_D) using Equation (2.63), one obtains a non-linear equation involving the Reynolds number Re = $v_{pb}d_{pb}/\nu$ as the independent variable. By solving the equivalent of Equation (2.62) numerically, one can calculate Re, and from this, determine v_{pb}. Applying this approach to data cited Fukushi *et al.* (1998) leads to the plot shown in Figure 3.30. The experimental data associated with this plot was gained from a 50 mg/l kaolin suspension coagulated with aluminium sulphate using a dose of 2.5 mg Al/l at pH 7.5. The resultant flocs are characterised by an ALT ratio of 1: 20.

Reference to Figure 3.12 indicates that the flocs may be represented by $A' \approx 6.5$ and $D \approx 1.73$ when $d_o = 3.5$ μm and $\rho_s = 2650$ kg m^{-3}. The volume mean value of d_b is cited as 63 μm. Data in Fukushi *et al.* (1998) indicates that n_b is roughly proportional to d_p^2. Physically this might be expected on the grounds that the number trapped should depend on the available surface area. The 'parking area' of a bubble is scaled by d_b^2. Hence one might expect n_{pb} to comply with the expression:

$$n_{pb} = k\left(\frac{d_p}{d_b}\right)^2 \tag{3.35}$$

Whatever its precise form, n_b is likely to depend on the size ratio. The values $k \approx 0.7$ and $d_b = 63$ μm were implemented in the stated calculation scheme. Overall, the trend shown in Figure 3.30 provides a good representation of experimental data reported in Figure 11 of Fukushi *et al.* (1998).

When applying the framework to small flocs (say < 50 μm), three problems are faced. First, is the acute lack of knowledge of the floc solids concentration in this size domain. Second there are representational difficulties when flocs possess a multifractal structure. Last, one must gain knowledge of association between n_{pb} and d_p.

Another facet of concern in DAF units is the impact of turbulence at the base of the reaction zone. Masschelein (1992) cites G values on the order of 1000 to 1500 s^{-1} as being 'maintained throughout the basin', but the origins of this information is unclear.

On the basis of floc breakage data, Fukushi *et al.* (1995) estimated the effective energy dissipation in the contact zone to be about 0.006 W cm^{-3} for a batch experiment and 0.06 W cm^{-3} for continuous flow. At 20°C, these are equivalent to G values of 2400 s^{-1} and 7600 s^{-1} respectively. Bache and

Rasool (2001a) note that other approaches have been used for such estimates, but generally information is lacking.

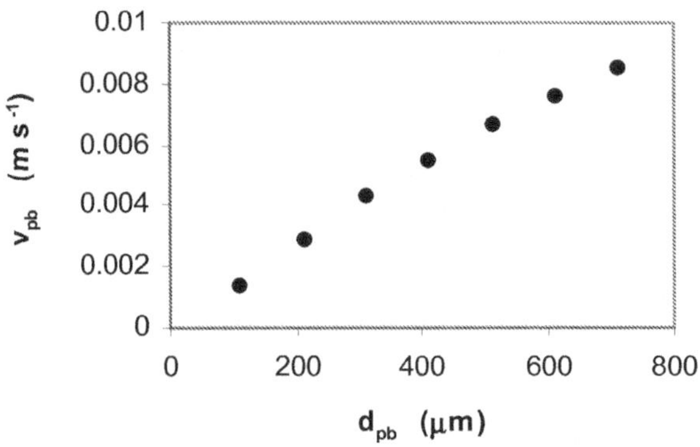

Figure 3.30. Calculated dependence of rise velocity as a function of floc size based on data cited in Fukushi *et al.* (1998) and Tambo and Watanabe (1979).

The consequences of high 'G values' is best understood in terms of the corresponding value of Kolmogorov length (η), this being ~ $(\nu/G)^{1/2}$. If G = 1000 s^{-1}, $\eta \approx 32$ μm. If one uses a rule of thumb that $d_{max} \sim \eta$ to 10η (Bache, 2004), this translates to a maximum sizes of 30 to 300 μm. Given that mean sizes are well below the maximum size (*e.g.* Table 3.11), one can see the consequences of having localised regimes of high energy dissipation.

Clearly there is little point in producing large flocs in flocculators only to break them in a contact unit. These views are illustrated in Bache and Rasool (2001a) for the case of alumino-humic flocs. On the other hand, if the data shown in Figure 3.30 is representative of the impact of floc size, one can see that separation is aided by larger flocs (by virtue of a larger area for bubble attachment and a lower effective density stemming from their fractal structure).

If one encounters excessive carryover (Table 3.17), this might be attributed to inadequacies in collection, short-comings in separation, or arising from hydraulic problems such as too much energy input. The importance of n_b (in collection) points to a requirement to form and sustain a relatively homogeneous bubble cloud for a sufficient time so that the bubble cloud fulfils its purpose. In practice, there are diminishing returns when the air dose exceeds about 8mg /m^3.

Table 3.17. Excessive carryover in DAF: problems and remedies.

Insufficiency of dissolved air supply
• Adjust either or both recycle rate and pressure to provide 6-8 g/m^3 dose of air • If air dose of more than about 8-10 g/m^3 needed, check function of or change the injection nozzles • Assess whether recycle is dispersed efficiently after injection and air comes out of solution quickly. • If premature air release appears to be occurring in delivery lines then reduce nozzle size or number of nozzles used
Large air bubbles appearing at head of DAF tank
• Check pressure before recycle injection is high enough to prevent premature air release • Check that no nozzles have been damaged.
Excess fine turbidity
• Check effectiveness of coagulation, flash mixing and flocculation • Consider whether recycle injection smashes floc: alter injection geometry, reduce injection pressure (using larger or more nozzles) or perhaps use a polyelectrolyte
Excessive surface loading
• Check sensitivity of carryover to flow rate. If flow for acceptable quality is lower than expected, consider hydrodynamic/hydraulic modifications to alter flow and/or retention time
Treated water quality is poorer than expected when the water is coldest
• Check choice of coagulant: e.g. change from alum to PACl, and its optimum dose • Check coagulation pH: optimise for pOH • Check and possibly improve dispersion of chemicals • Try using a polyelectrolyte.

3.3.6 Filtration

3.3.6.1 Foreword

Perhaps the most important attribute of filtration is the ability to deliver filtrate of an acceptable water quality. Behind this there are issues of net plant production and production efficiency. In deep bed filtration, vulnerable points in the filtration cycle occur immediately after back washing as filters are brought back on line and at the later stage due to the exhaustion of the removal capacity of a dirty filter. Breakthrough after back washing can be managed by bringing filters on line more slowly (*e.g.* Hudson, 1981). Implications of the incoming particle size were briefly noted in Section 3.3.1.2. For *favourable* filtration (a term used in O'Melia, 1985), the feed particles must be able to adhere to the collection surfaces *i.e.* 'sticky'. This can arise when the particles and collectors have opposite charges as in the filtration of positively charged metal hydroxides by negatively charged sand at a pH of 6. In unfavourable filtration, the existence of an energy barrier of a few kT can affect attachment (Section 1.2.10). The inclusion of a polymer prior to filtration can help to promote adhesion. Here the intention is to examine factors that affect the solids handling capacity and breakthrough.

In addition to the goals on filtrate quality, restrictions are imposed by the maximum headloss, which can be sustained across the filter bed, this normally being limited to 2.5 to 3 m. In laminar flow conditions, the head loss (h) across a uniform bed of depth L can be evaluated using the Kozeny equation (*e.g.* Cleasby, 1990):

$$\frac{h}{L} = \frac{k_f \mu}{\rho_w g} \frac{(1-\varepsilon)^2}{\varepsilon^3} \left(\frac{a}{\mathrm{v}}\right)^2 u \tag{3.36}$$

in which ε is the porosity, u the superficial velocity (= flow rate/bed area) and k_f, a constant close to 5. The ratio a/v is the surface area per grain volume, which takes the value $6/d_g$ for spherical grains and $6/(\psi d_g)$ for irregular grains with d_g as a grain diameter and ψ a shape factor. The Kozeny equation is applicable in the regime Re < 3 with Re = ud_g/ν. In constant rate filtration, it is seen that the head loss is controlled by the porosity. As filtration progresses, the porosity is reduced as a result of deposition, ε being defined by:

$$\varepsilon = \varepsilon_o - \sigma_v \tag{3.37}$$

in which ε_o is the porosity of a clean bed and σ_v, the specific deposit (= total volume of deposit/bed volume). For a bed made up of rounded or worn sand ε_o is about 0.42.

As filtration progresses, the presence of the deposit alters the effective surface area available for deposition, thereby affecting the trapping efficiency. Though there are many models for describing the action of filters, the conceptual approach described in Ives (1975) brings together some of the principal features. In this approach, the removal across a layer of thickness δL is expressed by:

$$\frac{dC_v}{dL} = -\lambda C_v \tag{3.38}$$

in which C_v is the volume concentration (vol/vol) of the suspension and λ the filter coefficient. The filter coefficient is a function of the deposit *i.e.* $\lambda = f(\sigma_v)$. Its principal features are expressed by the Ives model:

$$\lambda = \lambda_o + a_1\sigma_v - a_2\frac{\sigma_v^2}{\varepsilon_o - \sigma_v} \tag{3.39}$$

in which a_1 and a_2 are fitting coefficients. When a bed is clean $\lambda = \lambda_o$. As filtration progresses, λ first increases as a result of increasing contact area, passes though a maximum and then declines as the available trapping area is reduced, eventually reaching the limiting condition $\lambda \approx 0$. At this limit there is no removal across the layer, implying that the entire solids feed gets transmitted to the layer below. When $\lambda \approx 0$, there is a balance between the deposition rate and the rate at which the deposit is stripped away by shear forces. At this condition, which is also expressed by $d\sigma_v/dt = 0$, the prevailing value of σ_v is denoted by the symbol σ_{vu}. As the filter progresses towards this state, the headloss will continue to increase and may breach its limiting value. Alternatively, one may find that the renovation capacity of the filter becomes exhausted, this being shown by a deterioration in the filtrate quality. When the filtrate concentration exceeds some critical limit, the situation is referred to as *filter breakthrough.*

3.3.6.2 *Impacts of size*

Although conceptually useful, Equations (3.38) and (3.39) do not provide direct insight into the implications of the impact of particle size, this applying to both the incoming particles and the grain size (d_g) of the filter bed. A framework for describing this aspect is embodied in Equation (3.19), which is often referred to as the 'single collector' model. The logic which applies is similar to that described in Section 3.3.5. For example, Yao *et al.* (1971) envisaged a filter as being comprised of slices of thickness δz, within which the removal can be expressed by:

$$\frac{dn}{dz} = -\frac{3}{2}\frac{(1-\varepsilon)\alpha_{pc}\eta_{pc}}{d_g} \tag{3.40}$$

The notation used in Equation (3.40) is the same used in Equation (3.19) with ε referring to the prevailing porosity. From this, the collection efficiency across a filter slab of depth L is expressed by the relationship:

$$E_c = 1 - \exp\left[-\left(\frac{d_g}{L_{char}}\right)\left(\frac{L}{d_g}\right)\right] \tag{3.41}$$

in which L_{char} refers to a characteristic removal length defined by:

$$L_{char} = \frac{2}{3}\frac{d_g}{<(1-\varepsilon)\alpha_{pc}\eta_{pc}>} \tag{3.42}$$

The term <...> refers to a representative value of the enclosed term across the filter depth. In Montgomery (1985), the authors proposed that the ratio of the filter depth to media size (L/d_g) be used as a principal design variable – a common specification being $L/d_g \approx 1000$. After this factor has been fixed (by the design), it is evident that the removal efficiency depends on the ratio L_{char}/d_g (Lawson and Nason, 2006), this being controlled by the factors shown in Equation (3.41). With good coagulation practices, the attachment efficiency attains the value $\alpha_{pc} = 1$ *i.e.* "sticky". The term η_{pc} follows the schematic trend shown in Figure 3.26, further detail being provided in Lawler and Nason (2006). Its value depends on the size and density of the incoming stream of particles. Thus L_{char} does not have a unique value because of this dependence. Lawler and Nason argued that it was useful to assess filter performance in terms of minimum removal efficiency (based on the lowest value of η_{pc}) as a means of circumventing the size dependence - the logic being that if one can remove the 'most difficult' sizes (say ~ 1 μm), remaining particle sizes will be removed more efficiently.

It should be emphasised that Equation (3.41) applies to changes taking place across a thin slice. The identification of L with the filter depth (as distinct from its association with thin slices) can only be applied in situations in which the controlling parameters (such as appearing in Equation 3.42) have relatively small spatial variations within the slice. In deep bed filtration, the specific deposit (σ_v; affecting ε) can display strong spatial variations as illustrated in Hudson (1985). In this type of situation, one must examine the collection process slice by slice. These can be integrated across the entire filter depth using a modelling scheme of the type described in Bache (1984).

3.3.6.3 Solids handling capacity

Adin and Rebhun (1977) developed a model for evaluating and predicting filter performance, paying particular attention to the impact of water quality parameters, coagulants and the filtration rate on features of the deposit. Of particular interest is their use of a parameter (F) defined as the *theoretical filter capacity* (the mass of flocculated material per bed volume which would clog the pores completely). The value F was determined experimentally and involved extrapolation of a relationship between a headloss ratio and the specific deposit. Values of F are summarised in Table 3.18.

In the experiments conducted by Adin and Rebhun, the lowest mass capacity was obtained for humic acid with alum, and the highest capacity for kaolin with polymer. Differences in F were attributed to the differences in the deposit density. The notional solids concentration (c) (mass/ volume of deposit) is determined by the identity:

$$c = \frac{F}{\varepsilon_o} \tag{3.43}$$

and the corresponding deposit density specified by:

$$\rho_d = \rho_w + c\left(1 - \frac{\rho_w}{\rho_s}\right) \tag{3.44}$$

In Table 3.18 the dry density of the solids (ρ_s) was taken as 2500 kg m^{-3}. Rebhun *et al.* (1984) remark that the order of increase in solids concentration of the deposit and the density are consistent with expectation, remarking that alumino-humic sludges without mineral content have a relatively low solids concentration and density, whereas a clay mineral-polymer floc sludge has a relatively high solids concentration and density. The salient point to emerge is that the character of the deposit reflects the character of the source floc. This is significant, because it is the solids content of the deposit which influences its strength (Section 2.3.5) and thereby its susceptibility to shear.

As filtration progresses Ives (1975) envisages a situation in which side spaces are filled in and flow confined to channels approximating tubes or capillaries. Using such a model, Ives shows that the [coating] thickness of a deposit is specified by:

$$\theta = r\left[1 - \left(1 - \frac{\sigma_v}{\varepsilon_o}\right)^{1/2}\right] \tag{3.45}$$

in which r is capillary radius ($\cong d_g/2$). Given that the source material has a fractal structure, it is inevitable that the deposit will share this feature. On this basis, its solids content is likely to be expressed by Equation (2.22) in the form $c = A' \rho_s(\theta/d_o)^{D-3}$. Given this assumption, the solids concentration (mass held in deposit per unit bed volume) might be estimated using:

$$F_\theta = c\sigma_v = A'\rho_s\left(\frac{\theta}{d_o}\right)^{D-3}\varepsilon_o\left[1-\frac{(r-\theta)^2}{r^2}\right] \quad (3.46)$$

When $\theta = r$, $F_\theta = F$ (as defined in Rebhun *et al.*, 1984). The value of this exercise is that the parameters included in the definition of '*c*' convey information about the character source floc from which the deposit is formed. Lacking detailed information, a first order estimate of c can be gained by assuming that it is similar to the source floc. As an example, the alum-clay deposit might be characterised by the values $A' = 3.7$, $D = 1.9$, $d_o = 3.5$ µm and an alumino-humic deposit by $A' = 2.3\text{x}10^6$, $D = 1.15$, $d_o = 0.008$ µm. Referring to Table 3.18, the bed characteristics can be specified by $\varepsilon_o = 0.4$, $r = d_g/2 = 605$ µm. With this information, characteristics of the deposit are summarised in Table 3.19.

Table 3.18. Theoretical maximum filter capacity in terms of mass per unit bed volume, calculated solid concentration per unit deposit volume and associated density of deposit in bed (Rebhun *et al.*, 1984).

Suspension	Grain Size (mm)	Bed porosity ε_o	Maximum mass capacity $F=c\varepsilon_o$ $\left(\frac{\text{mg}}{\text{cm}^3}\frac{\text{solid}}{\text{bed}}\right)$	Solids conc. in deposit c $\left(\frac{\text{mg}}{\text{cm}^3}\frac{\text{solid}}{\text{deposit}}\right)$	Density of accumulated deposit in bed ρ_d (g cm^{-3})
Humic acid + alum	1.21	0.40	08.1	20.3	1.012
Humic acid + kaolinite + alum	0.62	0.37	13.7	37.0	1.022
Humic acid + alum + polymer	0.62	0.37	16.0	43.2	1.025
Kaolinite + alum	1.21	0.40	25.5	64.0	1.038
Kaolinite + alum	0.61	0.37	30.2	82.0	1.049
Kaolinite + polymer	1.21	0.40	48.0	120	1.072

Several features are evident. First, if one compares the values of c in Tables 3.18 and 3.19 for similar suspension types at the condition $\sigma_v/\varepsilon_o = 1$, it is evident that c is underestimated - indicating a denser solids packing within the deposit. Second, and of greater significance, it is apparent that the mass handling capacity (indicated by F_θ) is favoured by thinner deposits. Put another way, the thicker deposit offers no advantage in terms of mass accumulation. This behaviour stems directly from the [assumed] fractal behaviour. Behind this is a message that filters will work more efficiently (in terms of mass trapping) when supplied with smaller flocs of higher solids concentration. The principal demand on size is that they should be sufficiently large (*e.g.* a few μm) to be trapped efficiently.

Table 3.19. Illustrating impact of deposit thickness on solid handling capacity of a filter bed.

Suspension type	θ (μm)	σ_v	c $\left(\frac{\text{mg solid}}{\text{cm}^3 \text{ deposit}}\right)$	$F_\theta = c\sigma_v$ $\left(\frac{\text{mg solid}}{\text{cm}^3 \text{ bed}}\right)$	$\frac{\sigma_v}{\varepsilon_o}$
Alumino-humic	50	0.06	500	32	0.16
	100	0.12	137	17	0.30
	605	0.4	5	2	1.00
Kaolinite-alum	50	0.06	525	33	0.16
	100	0.12	245	30	0.30
	605	0.4	34	14	1.00

3.3.6.4 Deposit stability

Referring to Bai and Tien (1997), the shear stress acting on a collector in a filter bed may be estimated using the Happel model:

$$\sigma = 3\mu \frac{A_s}{d_g} u_f \tag{3.47}$$

where A_s is defined as $2(1-p^5)/w$ where $w = 2 - 3p + 3p^5 - 2p^6$, $p = (1 - \varepsilon)^{1/3}$ and u_f is the local approach velocity. In a packed bed with filter grains, the approach velocity may be approximated by the average interstitial velocity of the bed *i.e.*

$$u_f = \frac{u}{\varepsilon} = \frac{u}{\varepsilon_o - \sigma_v} \tag{3.48}$$

in which u is the filtration rate. Equation (3.47) has been applied to experimental data cited in Adin and Rebhun (1977) as a means of generating information about the applied stresses acting on the deposit. The study centred on the filtration behaviour of a coagulated clay suspension through a column of sand. Relevant data is summarised in Table 3.20.

Table 3.20. Practical and theoretical capacity of a filter bed together with estimated shear stress acting on ultimate specific deposit (based on data in Adin and Rebhun, 1977).

d_g (mm)	ε_o	Coagulant	Filter rate u (m/h)	Theoretical Filter Capacity F (mg/cm^3)	Practical Filter Capacity σ_{mu} (mg/cm^3)	Ultimate Specific Deposit σ_{vu} (vol/vol)	Shear stress σ (N/m^2)
1.21	0.4	polymer	10	48	20.1	0.17	4.1
1.21	0.4	polymer	15	69	35.5	0.21	11.4
1.21	0.4	polymer	20	71	30.3	0.17	8.3
1.21	0.4	polymer	25	82	27.3	0.13	6.1
1.21	0.4	alum	10	26	7.1	0.10	1.7
1.21	0.4	alum	15	21	4.6	0.08	2.1
1.21	0.4	alum	20	37	5.6	0.06	2.3
1.21	0.4	alum	25	20	2.8	0.055	2.7
0.62	0.37	alum	5	30	11.7	0.14	4.0
0.62	0.37	alum	10	33	7.2	0.08	3.8
0.62	0.37	alum	14.7	30	4.6	0.06	4.5
0.62	0.37	iron	7.35	10	3.6	0.14	5.9
0.62	0.37	iron	14.7	12	3.6	0.11	8.0
0.62	0.37	iron	14.7	15	3.4	0.09	6.3

The calculated shear stress is indicative of the yield stress that can be sustained by the deposit. Of particular interest is the relative strength of the deposits, rather than the absolute values of the calculated stress. For the coarser sand, it is seen that the polymer results in a deposit with greater shear strength than when using alum. In the case of the finer media, it appears that iron-clay deposit is stronger than that associated with alum. It is of interest to note that Hanson and Cleasby (1990) report that alum-kaolin flocs were significantly weaker than their iron equivalent. Viewed as a whole (including the effects of the polymer), it appears that the strength of the deposit reflects the 'characteristic strength' of the incoming floc.

An interesting example of the implications of deposit stability is illustrated in Figure 3.31; this refers to data (unpublished, RG) obtained from a pressure filter pilot plant operated in direct filtration at a rate of 4 m/h. Flocculation took place in the pipework before the filter. Headloss was measured through the pipework and used to calculate the average velocity gradient. The incoming stream was a low alkalinity coloured water coagulated with alum (dose ≈ 3.5 Al mg/l). It should be noted that flocs of this type are regarded as 'weak', strength deteriorating with increasing size (Section 2.4). Reference to Section 1.3.3.3 indicates that the product Gt (Camp Number) scales the extent of flocculation and thereby the opportunity for floc development; the likelihood is that flocs formed at low Gt values will be smaller than those formed at higher values. Figure 3.31(a) suggests that run times are greatest when the flocs are smallest. At the largest Gt, Figure 3.31(b) shows that the headloss development rate is low.

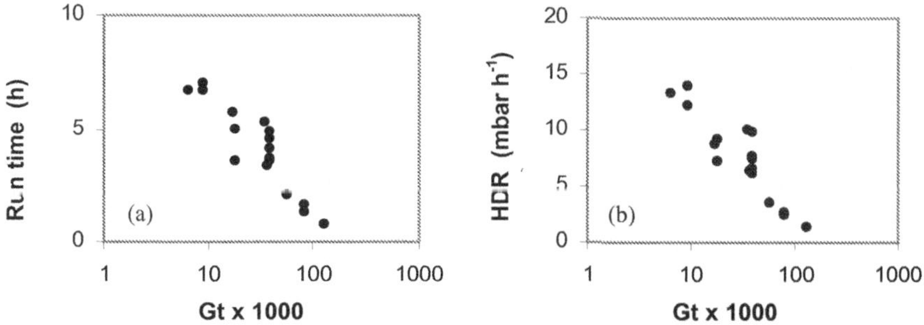

Figure 3.31. (a) Filter run time and (b) corresponding *headloss development rate* (HDR) in terms of product Gt x 1000 in which Gt is the Camp Number (unpublished data, RG).

The most likely explanation is that the associated [larger and weaker] flocs lead to the formation of an unstable deposit, which is readily scoured by the prevailing shear stresses. The resulting equilibrium deposit is likely to be thinner, thereby reducing the head loss. A further consequence is that the unattached solids will be rapidly transmitted through the filter, to emerge as breakthough at early times (see Figure 3.31a). It is evident that run times illustrated in Figure 3.31(a) are being controlled by breakthough, as distinct from the accumulated head loss. When Gt is low, the deposits are relatively strong and stable (indicated by the higher rate of headloss development and the longer run times). Although evidence is indirect, it points to intricate coupling of size and strength insofar as they influence filter performance.

3.3.6.5 Operation problems

A number of operational problems are noted in Tables 3.21 and 3.22. With regard to increases in headloss, many of the remedial measures advocated should be seen in the context of Equation (3.30). Increases in headloss are driven by the *volume* of the deposit. Although this is scaled by the incoming mass load, the volume is dominated by the associated water content. With respect to early breakthrough, this is a failure of solids retention. Such failures might be attributed to poor adhesion, unsuitable floc sizes or weak deposits. Again the association between water content and strength/weakness of flocs and deposits must be stressed. The analysis contained in Tables 3.18 – 3.20 rests on many assumptions, some of which are speculative. Nevertheless, it highlights features that are likely to affect filter performance. To repeat and emphasise the earlier statement of Rebhun *et al.*, (1984) "...low density deposits correlate with early breakthroughs, short runs and low filtration rates…".

Table 3.21. Operational problems associated with direct filtration.

Combating rapid rate of headloss

- First, check that polymer is not being over-dosed (to reduce floc size and/or stickiness) and check that the coagulant is not being over-dosed
- Reduce rate of filtration or increase the number of filters (velocity component of loading rate)
- Change the chemistry (to reduce the concentration component of loading rate) *e.g.* partially replace Al or Fe with a cationic polymer
- If feasible, use source water of better quality
- Use media with larger grain size (d_g) and/or use shallower (L) bed to reduce L/d_g value; but must avoid early breakthrough
- If headloss is localised at upper (approach) surface, then partially replace sand with anthracite (so reducing L/d_g value)

Addressing early breakthrough (when headloss is still low)

- Ensure that coagulation dose and pH are optimal
- Check that G value before entry to filter is not excessive and might be producing a non-reforming micro-floc; also check that Gt is adequate for macro-floc formation
- Use a polyelectrolyte (to increase strength of deposit), but avoid overdosing
- If problem associated with cold temperature, then seek to increase delay time to improve flocculation and/or change from alum to PACl.

Addressing media fouling (e.g. mud binding, organic debris)

- Check that backwash regime is effective and modify as necessary
- If fouling takes form of bed surface matting, consider replacement of sand by anthracite or introduce clarification
- If fouling takes form of in-depth 'mud' (congealed media, bed cracks, mud pellets), check backwash regime and/or stop excessive use of polyelectrolyte and then change filter sand
- If fouling takes form of in-depth 'slime' (from supracolloids?), check backwash regime and need to improve coagulation of DOC/NOM and increase delay time for cold water treatment
- If fouling takes form of sand size increase due to carbonate or manganese precipitation then change filter sand and review chemistry and processes needed.

Table 3.22. Operational problems associated with clarification-filtration.

Combating rapid rate of headloss

- Check chosen chemistry *e.g.* effectiveness of chemical dispersion, order of dosing, strength of chemicals dosed, use of carrier water, use of delay time and application of flocculation.
- Check not using excessive dose of polymer.
- Take account of seasonal variation in raw water quality and water temperature
- Improve clarified water quality (*e.g.* modification of flow retention characteristics or increase numbers of clarifiers, use of ballasting agent)

Addressing early breakthrough (when headloss is still low)

- Improve quality of floc for filter retention: actions similar to those described in Table 3.21

Addressing media fouling

- Not so likely to occur, but actions as listed in Table 3.21

3.3.7 Dewatering

3.3.7.1 Perspective

The prime purpose of dewatering is to increase the sludge solids fraction to achieve volume reduction. This of course is the endpoint of a sequence of processes in which suspended solids become more concentrated. A summary of the solids concentrations associated with different treatment stages is given in Table 3.23. The solids load can be estimated from knowledge of the flow and solids concentration in flocculators, but should also include additional inputs *e.g.* ballasting agents. Alternatively, estimates can be based on formulae such as Equation (1.58). When dosing polymers, knowledge of the solids concentration is vital. The nature of sludges mirrors the source floc, which depends on the composition of the raw water and coagulation practices.

Flocculated suspensions are highly non-Newtonian, except when they are very dilute. Even at the lowest solids concentration shown in Table 3.23 there is evidence of non-Newtonian behaviour. Background on this aspect is outlined in Section 2.3. Critical features include particle size, particle shape, the solids volume fraction and the electro-kinetic state. Of particular note is the gel point (ϕ_g), marking the state at which the solids within the suspension form a continuous network. When $\phi > \phi_g$ the network behaves as a solid. When sediments are subjected to compressive forces in this domain, dewatering depends on the strength and permeability of the network. Permeability also affects hindered settlement in the domain $\phi < \phi_g$ (*e.g.* Section 2.2.1).

Table 3.23. Solids concentration (outgoing) associated with treatment processes (Gregory and Dillon, 2002).

Process	Solids concentration (% w/w)
Floc blanket	0.2
DAF - scraped	1.2
- floated	0.4
Washwater - unsettled	0.03
- settled	1.0
Thickened - soft water	2.5-5.0
- hard water	5.0-10.0
Centrifuged - soft water	12.5-17.5
- hard water	17.5-22.5
Filter press - soft water	20.0-25.0
- hard water	25.0-35.0

It is useful to envisage dewatering as taking place in three stages: hindered settlement, [cake] filtration and expression. Thickening involves both settlement and expression during consolidation. Filtration is normally regarded as flow through a depositing cake and its supporting medium. Following deposition, subsequent releases (termed as *expression*) are the result of cake compression. During the expression phase, the applied pressure gradually transfers from liquid phase to the solids network *i.e.* with increasing the particle pressure and diminishing liquid pressure (see Figure 3.32). When flow ceases, the pressure difference across the cake (ΔP) is entirely supported by the cake solids. At this state, $\Delta P = P_y(\phi)$, $P_y(\phi)$ referring to the compressive yield stress of the cake or network.

Prior to discussing these stages, it is useful to provide background information on the factors which affect dewatering.

3.3.7.2 Water state

Before the commencement of flow, energy must be expended to overcome binding forces. In the case of pure water (free from the influence of surfaces) cohesive forces between water molecules are extremely low and can be neglected in practice. At the other extreme, water may be strongly adsorbed at a surface (Section 2.1.5) - requiring a considerable expenditure of energy to overcome the binding forces. Between these extremes, the forces acting on water should be regarded as being part of a continuum as demonstrated in Chen *et al.* (1987). The energy state of water molecules is affected by dissolved substances, their impact being encapsulated by the osmotic pressure. Pure water (at standard temperature and pressure; a reference state) is defined as having zero osmotic pressure.

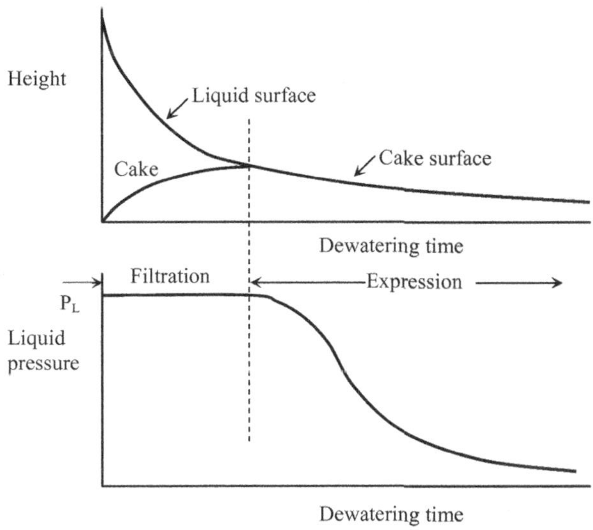

Figure 3.32. Phases of dewatering as indicated by cake depth and liquid pressure (Novak *et al.*, 1999).

When dissolved materials are present, osmotic pressures are always negative – indicating that dissolved substances cause water molecules to be held in a state of tension. The various forces acting on water *i.e.* gravitational, direct pressure, osmotic, colloidal each affect its energy state. These can be specified individually and combined additively to yield the total water potential (ψ) at any location. The potential (an expression of potential energy) is a measure of the work expended in transferring a water molecule from an arbitrary reference state to a particular location. It is often expressed in the form of energy per unit volume ($J\ m^{-3}$ ≡ thermodynamic pressure) or energy per unit mass ($J\ kg^{-1}$). Flow (*i.e.* Q) can only take place from points of high potential to points of low potential. Schematically $Q \propto \Delta\psi/R$ in which $\Delta\psi$ is the potential difference and R a medium resistance which restricts the flow. Readers may recognise this as being analogous to Ohm's law in classical physics.

When dewatering aids are added to a sludge, these can alter either or both the underlying tension forces and the resistive forces. For example, when organic polymers are used, their principal impact is on the rate of dewatering *i.e.* altering the resistive component, rather than affecting the tension forces.

Much of dewatering theory is expressed in terms of the solids volume concentration, whereas day-to-day measurements are normally expressed in terms of moisture content. The connection between the two is shown below.

For a sludge of mass (m_{sl}) containing solids of mass (m_s) and water mass (m_w), the moisture content (θ) is defined by the ratio m_w/m_{sl} and the solids fraction ($1-\theta$) by m_s/m_{sl}. From these definitions the moisture content can be expressed in the form:

$$\frac{m_w}{m_s} = \frac{\theta}{1-\theta} \tag{3.49}$$

Recognising that the mass of solids is conserved during dewatering, the interdependence between the sludge volume (V_{sl}) and the moisture content is expressed by:

$$V_{sl} = \frac{m_s}{\rho_s}\left(1 + \frac{\theta}{1-\theta}\cdot\frac{\rho_s}{\rho_w}\right) \tag{3.50}$$

in which ρ_s and ρ_w refer to the density of the solids and water respectively. When the system is described in terms of the solids volume fraction (ϕ), we note that $m_s = \phi V_{sl}\rho_s$ and $m_w = (1-\phi)V_{sl}\rho_w$ showing:

$$\frac{m_w}{m_s} = \frac{1-\phi}{\phi}\cdot\frac{\rho_w}{\rho_s} \tag{3.51}$$

Combination of Equations (3.49) and (3.51) shows:

$$\phi = \frac{\rho_w(1-\theta)}{\rho_w + \theta(\rho_s - \rho_w)} \tag{3.52}$$

An example of the application of Equation (3.52) is shown in Table 3.24.

Over decades, there has been a seemingly never-ending debate about what constitutes '*bound water*' (see also Section 2.1.5). In short, it depends how it is measured and the prevailing force fields causing water tension. In the context of an externally applied pressure, flow ceases when the pressure difference across the cake is entirely supported by the compressive yield stress of the network $P_y(\phi)$. At this equilibrium, the state of 'zero liquid pressure' (Figure 3.32 at long dewatering times) can be interpreted as a balance between the applied pressure (+) and the binding tension (-). This behaves in the same way as wringing a wet cloth or sponge. Examples of the association between the equilibrium moisture content and the applied pressure are shown in Figure 3.44. For present purposes, attention is drawn to the solids content at the highest pressure (10 bar), this being in the range 40-47% w/w. This is well in excess of the values attained in practice for this type of sludge (17-25% according to Table 3.23). The principal reason for the difference is the time factor, in that it can take several days before equilibrium conditions are reached for the sludge types illustrated in Figure 3.43.

Table 3.24. Example of association between moisture content (θ, %w/w) and solids volume fraction (ϕ, v/v) for the case $\rho_s = 2000$ kg m^{-3} based on Equation (3.52).

θ (%)	ϕ (v/v)	θ (%)	ϕ (v/v)
99.8	0.001	85	0.081
99	0.005	80	0.111
98	0.010	75	0.143
97	0.015	70	0.176
96	0.020	65	0.212
95	0.026	60	0.250
90	0.053	50	0.333

3.3.7.3 Impacts of coagulation

One of the distinguishing features of coagulation sludges is their moisture content - a factor singled out by Knocke *et al.* (1987) as affecting their susceptibility to dewatering. The conclusion was based on investigation into the dewatering behaviour of a kaolin suspension, which was coagulated with alum under differing modes of destabilisation. Significant results are summarised in Tables 3.27-3.28 (discussed later). The study showed that destabilisation by enmeshment led to the formation of large flocs of high water content and poor dewatering characteristics, whereas destabilisation by adsorption-charge neutralisation (ACN), yielded smaller and denser flocs with better dewatering characteristics.

In a subsequent investigation (Dulin and Knocke, 1989) the authors included the impact of organic matter on their model suspension. It was found that the presence of NOM increased the coagulant demand, causing more sludge, greater water content and a deterioration in dewaterability. In the 1987 paper, deteriorations in dewaterability were attributed to increased water content. While this is essentially correct, it centres on a symptom rather than the cause.

The key difference between charge neutralisation and enmeshment lies in relative abundance of bulk precipitate. When destabilising by enmeshment, there are copious quantities of amorphous hydroxide, whereas with ACN, amorphous hydroxide is still present, but in smaller quantities; its presence is largely inevitable because of the rapidity of hydrolysis compared with mixing timescales (Bache and Papavasilopoulos, 2003). Although NOM has hydrophilic attributes, NOM also affects the water content through its impact on coagulant demand. In Section 1.5.7 it was pointed out that precipitates have a great affinity for water - arising from mechanical trapping and hydrogen bonding. On a quantitative basis alone, one suspects that it is the relative abundance of coagulant products (particularly the amorphous hydroxides) that lie behind the observations of Knocke *et al.* (1987) and Dulin and Knocke (1989).

Beside their impact on water content, coagulation conditions also affect the rheological properties of a suspension. This arises from particle interactions under the influence of changing zeta potential (ξ) - this affecting the shear yield stress (Section 2.3.5.2). Briefly, when interacting particles are electrically neutral ($\zeta = 0$), van der Waals forces dominate, thus enhancing yield stress. However, when ζ becomes sufficiently large, repulsive forces overwhelm the attractive forces, leading to zero yield stress and a fluid or disperse state (features illustrated in Firth (1976) and Johnson *et al.*, 1999).

Figure 1.8 illustrates the impact of dose on the zeta potential of four suspensions coagulated with aluminium sulphate. With increasing dose, adsorption leads to the formation of a coating layer. In a state of partial coverage ($\zeta < 0$), the zeta potential is determined by both the deposit and charge state of the primary particles. In the domain $\zeta > 0$, the zeta potential attains roughly the same value for all suspensions. This stems from complete coverage by precipitate, the zeta potential being governed by the surface precipitate.

Aside from the impact of electrostatic bridging (Section 1.2.9), particle interactions will be dominated by the electro-kinetic state of the coating layer. The degree of coating affects the response of the system to pH. At very low doses, the ζ-pH trend is dominated by the original surface, whereas when the coating is complete, the ζ-pH pattern is governed by the surface deposit. An example of the impact of electro-kinetic factors on the yield stress of a concentrated kaolin suspension coagulated with alum is described in Johnson *et al.* (1999). The interplay of the zeta potential on the edges and faces of clay platelets provides a more complex yield stress–pH response than that associated with suspensions formed from spherical particles of uniform surface charge.

3.3.7.4 Permeability

At low Reynolds number, fluid transfer through a porous cake complies with Darcy's law (Equation 2.36). Within this, the most wide-ranging parameter controlling the transfer rate is the permeability (Section 2.2.2). When a coagulation sludge is formed from a model particle suspension *e.g.* kaolin, its composition is likely to be of the form illustrated in Figure 1.9, with clay particles forming a large-scale structure, the interstices filled with a hydroxide gel.

Guidance on the magnitude of the effective permeability can be gleaned from the specific resistance of a filter cake. Wakeman (1981) notes that the specific resistance (R_c) is consistent with the definition:

$$R_c = \frac{1}{k\rho_s(1-\varepsilon_c)} = \frac{1}{k\rho_s\phi_c} \tag{3.53}$$

in which the porosity (ε_c) and solids volume fraction (ϕ_c) refer to the cake. Permeability (k) can be specified by the framework described in Section 2.2.2. For ease of computation, we make use of the Happel model reported in Probstein (1995) *i.e.*

$$\frac{k}{d^2} = \frac{3-(9/2)\phi^{1/3}+(9/2)\phi^{5/3}-3\phi^2}{18\phi(3+2\phi^{5/3})} \tag{3.54}$$

in which d is the diameter of solid particles forming the bed. Combining Equations (3.53) and (3.54) and noting that $\phi \equiv \phi_c$ shows:

$$d = \left[\frac{1}{\rho_s R_c}\cdot\frac{18(3+2\phi^{5/3})}{(3-(9/2)\phi^{1/3}+(9/2)\phi^{5/3}-3\phi^2}\right]^{1/2} \tag{3.55}$$

The difficulty with this approach (particularly when $\phi > \phi_g$) is that it treats a continuous network as system of discrete particles. Also it does not recognise the impact of fractal structure. Nevertheless, it provides a means of estimating the size scales associated with the specific resistance. To illustrate this theme, we consider a cake with moisture content of 75% (w/w). Treating the solids density as 2000 kg m^{-3}, Equation (3.52) shows $\phi = 0.143$. Reference values of R_c leads to the estimates of k and d summarised in Table 3.25.

Table 3.25. Association of permeability (k) and characteristic cluster size (d) with specific resistance (R_c) in a porous cake.

R_c (m kg^{-1})	k (nm^2)	d (nm)	$k^{1/2}$/d
10^{14}	35	19	0.3
10^{13}	350	60	0.3
10^{12}	3500	191	0.3
10^{11}	35000	603	0.3

When the specific resistance exceeds about 10^{13} m kg^{-1}, the spatial scales (*i.e.* $k^{1/2}$ and d) are on the order of tens of nm, whereas when the specific resistance is below 10^{12} m kg^{-1}, the length scales are on the order of a few 100 nm. Given that the cited values of specific resistance reflect the range of values encountered in practice, the analysis indicates that permeability is controlled by a regime of size well below typical floc sizes and often below the 'large' particles such as shown in Figure 1.9. This theme pursued further in Section 3.3.7.9.

Changes in permeability can be induced by any process that causes the formation of larger and denser clusters, or a more open network. In terms of modelling this process, it must be recognised that the solids volume concentration is conserved. If the original solids concentration is specified by ϕ_c and fractal clusters of diameter ξ and volume fraction $\phi_F(\xi)$ are formed, then Equations (2.49) and (2.15) show:

$$\phi_F(\xi) = \frac{\phi_c}{\phi_\xi} \qquad \text{with} \qquad \phi_\xi = A'\left(\frac{\xi}{d_o}\right)^{D-3} \tag{3.56), (3.57}$$

i.e.

$$\phi_F(\xi) = \frac{\phi_c}{A'}\left(\frac{\xi}{d_o}\right)^{3-D} \tag{3.58}$$

This shows the coupling between ϕ_F and ξ. When clusters are porous, the permeability of the system can be evaluated using the scheme described in Section 2.2.2. In the seemingly straightforward case of impermeable clusters, the association between permeability and specific resistance can be based on Equation (3.53) and (3.54) with the symbols ϕ and d replaced by ϕ_F and ξ respectively in Equation (3.55). Solution demands knowledge of the structural parameters appearing in Equation (3.57).

3.3.7.5 Polymers

For thickening and dewatering of coagulant sludges, polymers of low or medium charge density and high molecular weight (refer to Table 1.8) give the best performance. With iron and aluminium salts as coagulant, the sludge particles generally have a positive charge so that an anionic polymer is appropriate. Non-ionic polymers, such as polyacrylamide are also used. The large molecular weight offers greater scope for inter-particle bridging and also strengthens clusters, making them more resilient to shear.

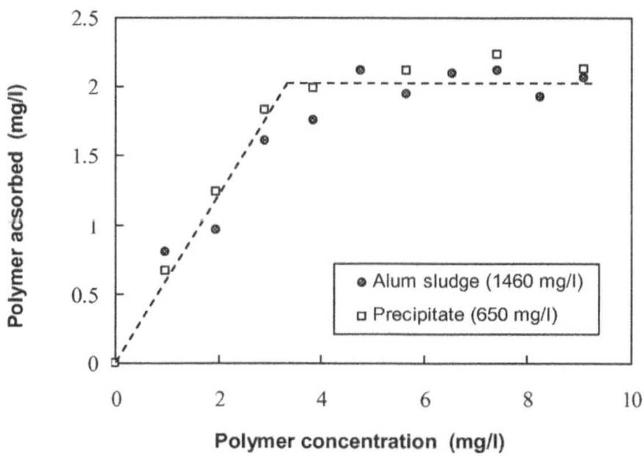

Figure 3.33. Adsorption of a non-ionic polymer on an alumino-humic sludge and bulk precipitate, the latter having solids concentration equivalent to its mass fraction within the sludge (Papavasilopoulos and Bache, 1998).

There is a vast literature dealing with interactions and the application of polymers to particle suspensions. Most of this applies to situations in which coagulants are not used. Hydroxides form the smallest scale structures in coagulation sludges and therefore are 'rate-determining' in dewatering. In this context, particular attention needs to be paid to polymer-hydroxide interactions.

Figure 3.33 shows a comparison of the adsorption of a non-ionic polymer on an alumino-humic sludge and on a freshly precipitated hydroxide suspension at a solids concentration equivalent to the coagulant fraction (≈ 45%) within the sludge. It is seen that adsorption increases with dose, saturation occurring at about 2.1 mg/g. However, the salient feature of Figure 3.33 is the coincidence of the trends. This shows that adsorption is dominated by the coagulant fraction. This is partly due to a lack of interaction between the humic materials and the particular polymer, but is also influenced by the much greater number of the low molecular weight hydroxide molecules compared with the high molecular weight humic molecules (Bache and Papavasilopoulos, 2003).

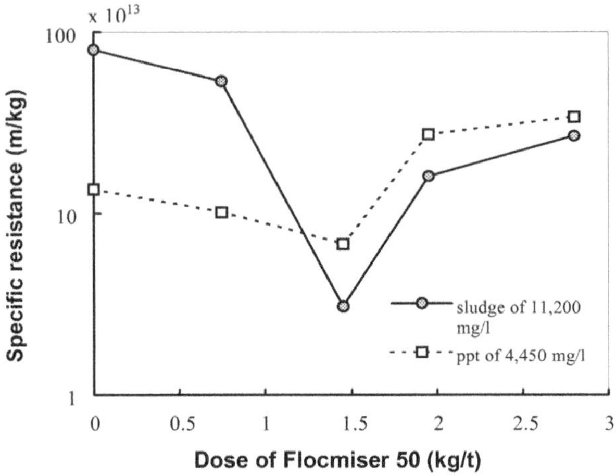

Figure 3.34. Impact of polymer dose on the specific resistance of an alum sludge and a suspension of precipitate (ppt) at the amount expected in the sludge (Bache and Papavasilopoulos, 2003).

A further example of the polymer-hydroxide interaction is shown in Figure 3.34, which compares the specific resistance (R_c) of an alumino-humic sludge with an equivalent hydroxide suspension. In both suspensions, the lowest specific resistance occurs at the same dose (1.45 kg t^{-1}), indicating that the minimum in the alumino-humic trend was being controlled by the polymer-hydroxide interaction. It is also evident that the specific resistance values are of the same *order* at corresponding dose.

The polymer-particle interactions occurring within a hydroxide gel differ greatly from those illustrated in Figures 1.6 and 1.7 because of the size and numeracy of the hydroxide particles. If saturation adsorption is specified as x kg t^{-1}, the number of hydroxide molecules of molecular weight M_{OH} associated with each polymer molecule of molecular weight M_p is given by:

$$\frac{n_{OH}}{n_P} = \frac{1000 M_P}{x M_{OH}} \tag{3.59}$$

For example, when $M_P = 10$ MDa, $M_{OH} = 78$ Da and $x = 1$ kg t^{-1}, $n_{OH}/n_P \sim 10^8$. Given these relative numbers, the hydroxides will rapidly saturate the binding sites. If the polymer is anionic and the hydroxides positively charged, one would expect the polymer coil to contract when its charges are neutralised. Equation (1.53) indicates that a polymer of molecular weight 10 MDa (*e.g.* a polyacrylamide) will have a coil size of about 230 nm. When the polymer and the hydroxides are combined, they might be envisaged as forming a cluster that behaves (in terms of permeability) as a large 'particle' with dimensions consistent with the coil size. An alternative view is that the polymer attaches itself to the surface of preformed flocs, thus binding them together (Wu *et al.*, 1997), however, what happens in practice is unclear.

3.2.7.6 Optima

One of the major sources of debate in dewatering is the optimum dose. Historically, guidance has been provided by trends in the Capillary Suction Time (CST) and specific resistance. Sometimes the information generated by differing test methods is contradictory, as was evident in Wu *et al.* (1997). In practice, one must be guided by the technical optimum *i.e.* what works best at full scale. Fundamentally, the magnitude of the optimum dose is steered by the adsorption chemistry. Saturation adsorption is usually expressed in the form mg g^{-1} or kg t^{-1}, the ratio showing the underlying linkage between the dose and the solids concentration. When the dose exceeds the saturation capacity, excess polymer exists in the form of a residual.

When the optimum dose is based on dewatering behaviour, or some other features of the polymer-sludge interaction, it should be recognised that the optimum is specific to a particular test. The optimum is usually interpreted as the dose corresponding to a minimum in a parametric behaviour-dose plot. In many cases, one finds that the minimum is a consequence of 'opposing' modes behaviour, such as illustrated in Figure 3.35. For convenience, factors causing a reduction in the magnitude of a characteristic parameter with increasing dose are called *decrease drivers* whereas factors causing an increase in magnitude with increasing dose are called *increase drivers*. An indication of the principal drivers associated with some test methods is given in Table 3.26.

Given the variety of interactions, it is hardly surprising that they give different estimates of the 'optimum dose'. While there can be differences of the type reported in Wu *et al.*, (1997), there can also be a reasonable consistency as illustrated in Figure 3.36. Indeed, this ought to be the case, because behavioural attributes are tied to the underlying pattern of adsorption. Within reason, inconsistencies between the test methods may not be too significant as there are many operational factors that affect the delivered dose as noted in Bache (2002).

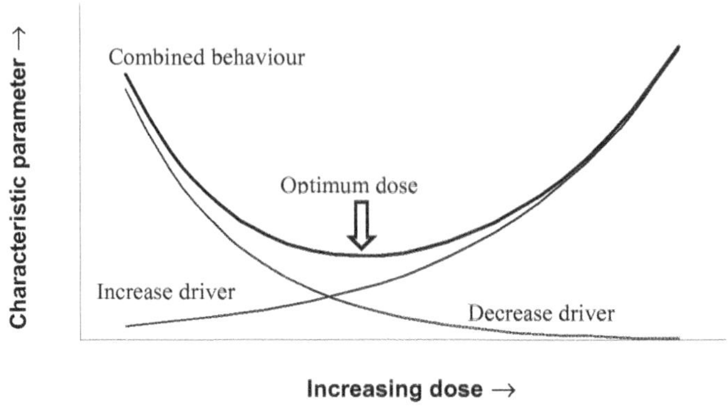

Figure 3.35. Schematic interaction of drivers that determine optimum dose.

Table 3.26. Summary of principal drivers that control the magnitude of optimum dose in standard test methods.

Test Method	Decrease driver	Increase driver	References
Centrate viscosity	Viscosity associated with turbidity	Viscosity due to residual polymer	Dentel & Abu-Orf (1995); Bache & Papavasilopoulos (2000)
CML30 (settlement)	Increasing cluster size and velocity	Behaviour of Settled bed	Bache and Zhao (2001)
Specific resistance	Increasing cluster size & permeability	Membrane resistance & excess polymer	Zhao *et al.* (1998)
CST	Increasing cluster size & permeability	Membrane resistance & excess polymer	Hayashi *et al.* (1990)

3.2.7.7 Shear

It is well known that the shear history affects the polymer dose. This has been documented in various studies of which Novak and Bandak (1997) is an example. The gist of their study was to examine the response of the CST to combinations of shear (represented by G), mixing time (t) and polymer dose in alum and lime sludges (from water softening).

When the alum sludge was unconditioned, it was found that CST $\propto G^{2.8}t$, showing a greater sensitivity to shear rather than time. Increasing shear will reduce the floc size and thereby reduce the effective permeability in the CST apparatus. When polymer was present, it was found that the CST-dose relationship passed through a minimum, the corresponding dosage depending on Gt. Optima were identified by minima, or the point at which CST < 10 s. It was found that the required polymer dose (*i.e* the optimum) increased with Gt. The precise reasons for this dependence are unclear, but presumably tied to the greater number of smaller flocs and larger surface area when shear conditions are increased. There is also the factor that when polymer linkages are broken, they do not reform, warranting additional polymer addition.

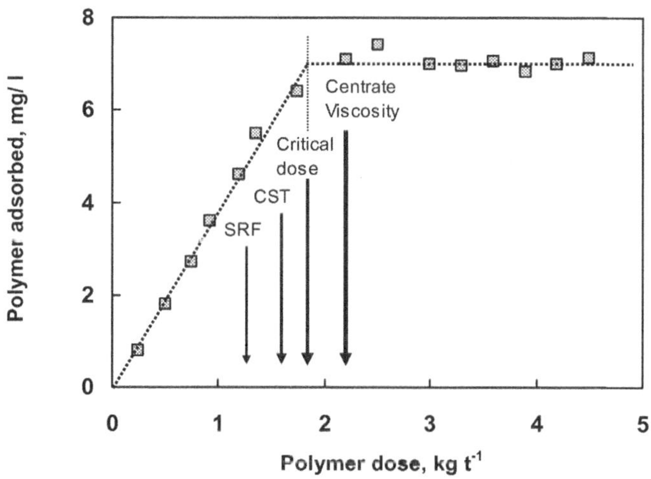

Figure 3.36. Comparison of optima identified by different test methods for an alumino-humic sludge. The 'critical dose' corresponds to the onset of adsorption saturation and SRF refers to specific resistance (Bache, 2002).

The implication of these findings is that the polymer dosage required for dewatering depends on the shear history. Thus, when selecting polymers and dosages for alum sludge conditioning, tests such as CST, should be carried out at shear levels resembling those encountered in the dewatering process.

In the case of the lime-softening sludge (made up primarily of $CaCO_3$), it was found that it did not deteriorate when subjected to shear, implying that the required polymer dose is relatively insensitive to the shear history.

3.3.7.8 Thickening

Thickening involves both hindered settlement and consolidation (Figure 3.37), both phases being affected by coagulation and conditioning practices. Before pursuing these in detail, it is useful to consider the underlying processes.

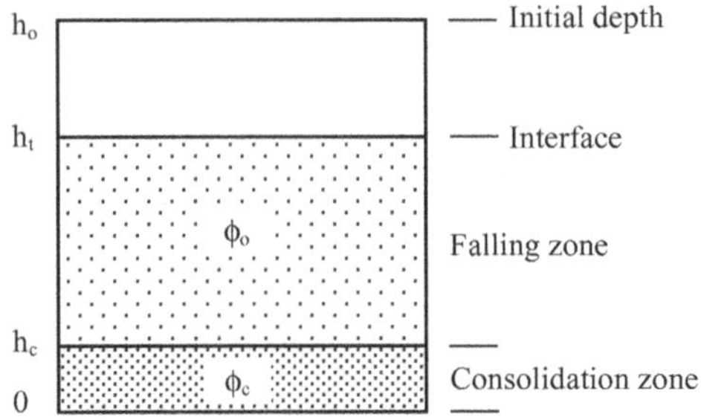

Figure 3.37. Zones associated with hindered settlement and cake filtration.

For suspensions with $\phi < \phi_g$ (gel point) flocs exist as discrete entities, their relative motion promoting collisions. In the early stages of batch settlement, the interface velocity increases with time as a result of growth. Increases in floc size imply a greater water content and greater susceptibility to self-induced shear (Section 2.4.2.3). Eventually a state is reached when the rates of growth and rupture balance, corresponding to a constant interface velocity. Hydrodynamic interactions between clusters cause them to move as a group. In principle this can be described by Equation (2.57a), permeability being tied to ϕ_F, but is complicated by the rheology. In practice, the steady interface velocity complies with Equation (3.21), this being a common means of characterising hindered settling. Many non-flocculating suspensions are represented by the values $k = 1$ and $p \approx 4.7$ in Equation (3.21) (the *Richardson-Zaki correlation*; see Probstein, 1994). In flocculated suspensions there are wide variations.

Four case studies are reviewed to illustrate different features of thickening. The first of these centres on Knocke *et al.* (1987), who examined the impact of the mode of destabilisation on dewatering of a kaolin suspension coagulated with alum. It was found that destabilisation by charge neutralisation produced sludges with far better thickening rates than those associated with enmeshment. These features were reflected in the moisture content and the settled volume of the sludge as shown in Table 3.27. The term AVI (*aggregated volume index*) is the ratio of the total floc volume to that fraction of the volume occupied by the dry sludge solids ($\equiv 1/\phi$); a guide to variations in water content. The superior thickening rates associated with ACN are directly attributable to the lower moisture content within the floc mass. For low turbidity waters (with enmeshment), it was found that both the interfacial velocity and degree of consolidation reduced with increasing pH. The reasons for the differences in water content are not clear, but are presumed to be caused by differences in electrokinetic behaviour (see case study below).

Table 3.27. Effects of coagulation mechanism on relative floc water content and water removed by gravity thickening (from Knocke *et al.*, 1987).

Coagulation conditions					
Influent turbidity NTU	Coagulant dose mg/l	pH	Mechanism	Ultimate thickened Solids concentration %	AVI
60	10	6.2	ACN*	6.0	90
40	10	6.3	ACN*	5.5	100
7	40	6.5	Mixed	1.0	700
7	75	7.1	Enmeshment	1.0	1060
7	75	8.1	Enmeshment	0.5	1600

*Adsorption – charge neutralisation

The second case study refers to Mpofu *et al.* (2003) who examined the influence of hydrolysable metal ions (Mn(II) and Ca(II)) and an anionic polyacrylamide-acrylate flocculant (PAM) on the dewatering of a kaolin suspension. Their study included measurements of zeta potential and Bingham yield stress. They found that there were very specific conditions favouring settlement – the optimum pH being tied to the appearance of the *first* hydrolysis products ($Mn(OH)^+$ and $Ca(OH)^+$) (strongly adsorbed on the particle surfaces). Metal ion - mediated zeta potential changes were reflected in lower dispersion yield stress and a higher settling rate. The adsorption of PAM was greatly improved by the prior addition of the metal ions. Optimum flocculation conditions involved partial rather than complete surface coverage by both metal ions and flocculant. Consolidation was also influenced by the coagulation conditions - higher yield stress dispersions displaying poorer consolidation. Above all, the study demonstrated the significance of interfacial chemistry on dewatering.

The third case study (Bache and Zhao, 2001) was designed to examine some of the implications of polymer overdosing on the dewatering of an alumino-humic sludge. Using a non-ionic polymer, the study yielded information about specific resistance and settlement/consolidation behaviour in response to solids concentration and polymer dose. Figure 3.38 shows detail of the response of interface height to polymer dose and time in a 1000 ml measuring cylinder (relatively free from wall effects). From the trends, it is evident that there is change in behaviour at about 10 mg/l irrespective of time. Above 10 mg/l, the falling phase is completed within 30 s, leading to the formation of a consolidated layer whose solid volume fraction is virtually independent of dose. This domain coincides with conditions of saturation adsorption. When this occurs (*i.e.* by over-dosing) consolidation is impaired. It is of interest to note that the solids volume concentration in settled layer is about 0.004; this being indicative of ϕ_g. Through time there is some consolidation, this being attributed to expression. Below 10 mg/l (≡ 2.2 kg/t), increasing dose enhances the rate of settlement, but also causes a deterioration in consolidation; this is most evident after prolonged settlement.

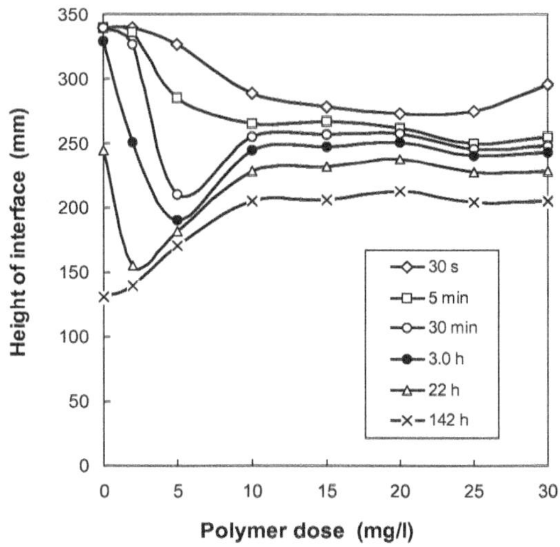

Figure 3.38. Response of interface height to dose and time for an alumino-humic sludge of 4595 mg/l solids concentration in a 1000 ml measuring cylinder (adapted from Bache and Zhao, 2001).

When one faces impacts on both the rate of settlement, and the degree of consolidation (such as illustrated in Figure 3.38), the question arises as to what is the best dose. In the study by Mpofu *et al.* (2003) noted above, the optimum conditions for clarification of a clay suspension did not coincide with best conditions for consolidation.

In the case of an alumino-humic sludge this led to the development of the *CML30 test* reported in Bache and Zhao (2001) and illustrated in Figure 3.39. This is an *ad hoc* test conducted in a *100 ml measuring cylinder* (with wall effects). It was found that the position of the interface after 30 minutes settlement provided a means of gauging the optimum dose (corresponding to the lowest interface height). The information generated was compared with optima identified by specific resistance testing. There was excellent agreement between the two approaches - giving credence to potential of the CML30 test as a means of identifying the optimum dosage for this type of sludge. Key advantages include the simplicity of the approach (useful for on-site investigations) and the fact that it responds to the sludge as found - whatever its solid concentration or shear history. Its potential use in practice is discussed below in the context of changes in solids concentration (Bache, 2002).

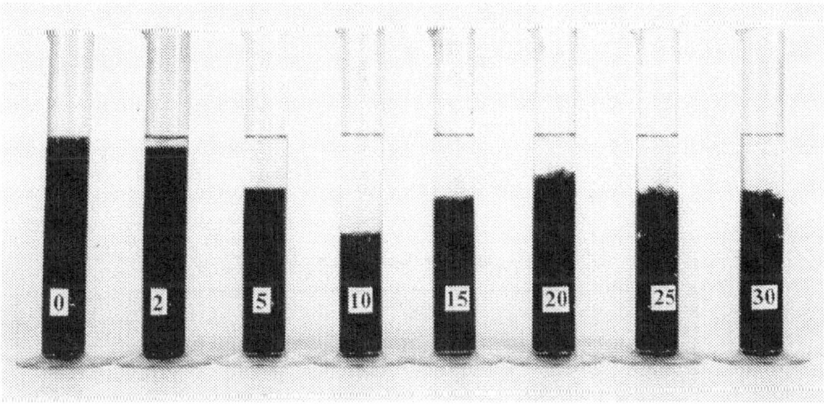

Figure 3.39. Response of interface height to dose (labelled in mg/l) at a settling time of 30 min settlement in a 100 ml measuring cylinder (CML30 Test). The lowest interface height (at 10 mg/l) is indicative of the optimum dose (Bache and Zhao, 2001).

Dosage is usually expressed in the form of the ratio kg (polymer)/tonne (dry solids). When one has a situation in which the dry solids concentration varies with time, it is necessary that the polymer feed rate responds accordingly in order to maintain the target dosage ratio. When this does not occur, *e.g.* maintaining a fixed feed rate, it is highly probable that both under-dosing and over-dosing will occur - this affecting the degree of consolidation. This aspect is illustrated in Figure 3.40, which refers to a waterworks in which the balance tank output was very variable and the polymer feed rate fixed. The settling behaviour of sludge samples was monitored using the CML30 approach. One sees that the settled volume responds to the changes occurring at full scale. In extreme circumstances (whether underdosing or overdosing) there is the danger of solids overflow as a result of poor settlement.

There are dangers in placing too much reliance on batch settlement tests of the type described above, because they do not reproduce the conditions found in a full-scale thickener. For example, slow stirring and raking enhances the thickening of all coagulant sludges, the effect being much greater after polymer dosing (Warden, 1983). Care should also be exercised about the potential effects of temperature through the impact on viscosity on the settlement rate (*e.g.* Equation 2.44 or via Equation 3.21).

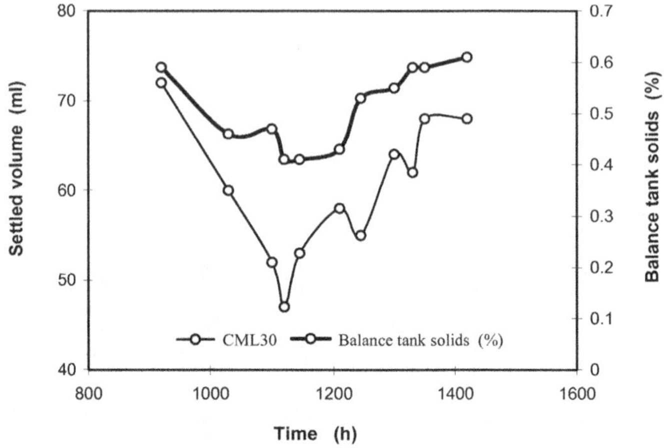

Figure 3.40. Relationship between solids concentration in a thickener and settleability in a CML30 at different times during a day (Bache, 2002).

3.3.7.9 Cake filtration

Specific resistance

Here we focus on flows through a depositing bed supported by some form of membrane. This situation is most often described by the framework developed by Coackley and Jones (1956) and since developed in various ways to take account of factors such as variations in bed structure (Wakeman, 1981). Deposits are normally regarded as fully networked ($\phi > \phi_g$).

In the approach described by Coackley and Jones, unidirectional filtration is represented by the model:

$$\frac{1}{A}\frac{dV}{dt} = \frac{\Delta P}{\mu R} \tag{3.60}$$

The left hand side is a superficial velocity based on the filtration rate dV/dt and the cross-sectional area (A), with V as the transmitted filtrate volume. Of the remaining parameters, ΔP is the pressure difference across the cake with μ as the dynamic viscosity and R the overall hydraulic resistance (units: m^{-1}) offered by the cake and its supporting membrane. By taking account of the pressure drop across the cake and the membrane, the hydraulic resistance can be split into components:

$$R = \frac{R_c m_s}{A} + R_m \tag{3.61}$$

in which R_c is the resistance per unit volume of the cake and R_m is resistance of the supporting membrane. The cake resistance is assumed to be proportional to the mass of solids deposited. In the form shown, R_c has units of m kg^{-1}. Treating the cake as uniform, $m_s/A=\phi_c\rho_s h_c$ in which ϕ_c is the solids volume fraction in a cake of depth h_c with ρ_s as the solids density. A liquid and solids balance across the regime shown schematically in Figure 3.37 yields:

$$h_c = \beta\frac{V}{A} \qquad \text{with} \quad \beta = \frac{\phi_o}{\phi_c - \phi_o} \qquad (3.62), (3.63)$$

With these inclusions, the filtration rate may be expressed as:

$$\frac{dV}{dt} = \frac{\Delta P A^2}{\mu R_c \phi_c \rho_s \beta V + R_m A} \tag{3.64}$$

When R_m is constant, integration of Equation (3.64) leads to the well-known dependence $V \propto t^{1/2}$ from which the specific resistance is determined. The association between specific resistance and the permeability is noted in Section 3.3.7.4. Recognising that cakes are compressible, this feature is often represented by the empirical expression:

$$\frac{R_c}{R_{co}} = \left(\frac{\Delta P}{\Delta P_o}\right)^S \tag{3.65}$$

in which ΔP_o is a reference pressure difference and R_{co} the corresponding value of the specific resistance, with S known as the *coefficient of compressibility*. When a cake is incompressible $S = 0$; otherwise, $S > 0$. When cakes are characterised, the effects of pressure should normally be included. Knocke and Wakeland (1983) report values of S in the range 0.79 to 0.97 for various types of alum sludge. When such sludges are dosed with polymer, Novak and O'Brien (1975) showed that S increased with polymer dose - the effects being most pronounced for an anionic polymer and least for a non-ionic polymer. S was found to be in the range 0.6 to 2.3. In extreme situations increases in pressure do not lead to increases in the filtration rate because of concomitant increases in specific resistance. Examples are shown in Novak *et al.*, 1999).

The investigations of Knocke *et al.* (1987) concerning the linkage between the mode of destabilisation and the dewatering of a kaolin suspension coagulated with alum included detail of the specific resistance and cake solids (Table 3.28). As in Table 3.27, one can see the association between the mode of destabilisation and cake characteristics, ACN yielding a cake of lower moisture content and specific resistance than destabilisation by enmeshment. Prior discussion (Section 3.3.7.3) drew attention to the differences in the hydroxide content, this being the likely cause of the findings reported in Knocke *et al.* (1987).

Table 3.28. Effects of coagulation mechanism on cake solids, specific resistance and effective particle size (based on Knocke et al., 1987).

Coagulation conditions					
Influent turbidity NTU	Coagulant dose mg/l	Coagulant pH	Mechanism	Cake solids % (w/w)	Specific resistance m/kg
60	10	6.2	ACN*	42	55 x 10^{11}
40	10	6.3	ACN*	22	60
7	40	6.5	Mixed	15	95
7	75	7.1	Enmeshment	11	150
7	75	8.1	Enmeshment	9	310 x 10^{11}

*Adsorption –charge neutralisation

Table 3.29 provides similar information for a wider range of sludge type. Although there are variations in the applied pressure difference used in the various studies (not shown), this aspect can be disregarded. The lowest specific resistance (~ 10^{11} m kg^{-1}) and highest cake solids are associated with the kaolin suspension when no hydrolysing coagulant is present. This contrasts with a suspension in which precipitates form 95% of the solid content, this having a specific resistance ~ 10^{15} m kg^{-1} and a high water content. The kaolin data is essentially a repeat of Table 3.28, but includes the impact of NOM. The information taken from Dulin and Knocke (1989) stems from a comparative study in which NOM was added. One can see that its presence caused a deterioration in dewaterability, but the impact is relatively small. This aspect was also scrutinised in Bache and Papavasilopoulos (2003) by varying the composition of the suspension, taking care to control the final cake solids (2%). The impact of humic substances was outweighed by the hydroxide fraction.

Table 3.29 also includes an estimate of the effective particle size associated with the specific resistance and solids content. The solids volume fraction was estimated using Equation (3.52) with due allowance for solids density. In the absence of precipitate the kaolin cake has an effective particle size close to 1 µm - this approximating the particle size of kaolin (a few µm). At the other extreme, the bulk precipitate has an effective particle size of a few nm - again matching the base particle size. In spite of the presence of NOM and polymer, one sees that the effective size reduces as the hydroxide fraction increases.

Table 3.29. Summary of specific resistance, final cake solids and effective particle size (using Equation 3.55) for kaolin and alumino-humic sludges under different treatment conditions.

Suspension	Features	Cake Solids % (w/w)	Specific Resistance m/kg	d nm	References
Kaolin + polymer	No alum	61	1.1×10^{11}	985	Besra et al. (2002)
Kaolin + alum	ACN*	42	5.5×10^{12}	87	Knocke et al (1987)
Kaolin + alum	Enmeshment	18	1.7×10^{13}	33	Dulin and Knocke (1989)
Kaolin + NOM + alum	Enmeshment + NOM	16	3.3×10^{13}	24	Dulin and Knocke (1989)
Alumino-humic + polymer	Polymer impact	1.7	1.0×10^{13}	36	Bache and Zhao (2001)
Alumino-humic	Mass ratio 75% humic 25% ppt*	2.0	6.0×10^{14}	4.9	Bache and Papavasilopoulos (2003)
Alumino-humic	Mass ratio 5% humic 95% ppt	2.0	8.5×10^{14}	3.5	Bache and Papavasilopoulos (2003)

*ACN ≡ Adsorption-charge neutralisation; ppt ≡ bulk precipitate

Overall, one can see that dewatering characteristics of the cake are sensitive to the hydroxide content. As the hydroxide fraction increases, the controls on dewatering become dominated by small scale behaviour, this being associated with the coagulant products.

Because the identity of specific resistance is tied to the phase of cake formation, the values of the cake solids shown in Tables 3.28 and 3.29 represent a guide as to the moisture content that can be attained *by filtration*. If one compares the solids content for and alumino-humic sludge in Table 3.29 with the corresponding information ("Filter press - soft water") in Table 3.23, it is evident that solids concentration obtained by filtration falls well below the values attained in practice. The difference can be attributed to the impact of the expression (Section 3.3.7.10).

Polymer overdosing

Earlier sections have drawn attention to the detrimental effects of overdosing on settleability, consolidation, and compressibility. Here we focus on the problem of filter blinding, this being associated with the *supra-colloids* (just above colloidal range). These can migrate through a cake and become lodged in critical passages or within the filter septum – especially particles/clusters with diameters in the size range 1 – 10 μm (*e.g.* Karr and Keinath (1978); Novak *et al.* (1988)). When this occurs it increases the flow resistance. An example of this feature is shown in Figure 3.41.

With increasing polymer dose, the flow resistance first decreases (due to increased permeability) and then increases because of filter blinding. In strict terms, the labelling on the vertical axis in Figure 3.41 is incorrect because it only applies to the domain in which the membrane resistance is constant *i.e.* before the emergence of the residual. Once the residual appears, the cake resistance remains low, the hydraulic resistance increasing as a result of the impact of the polymer residual. It reflects the behaviour of the membrane and *is not* a property of the cake. Zhao *et al.* (1998) found that the membrane resistance (R_m) complied with the empirical expression:

$$R_m = \alpha e^{\beta V} \tag{3.66}$$

in which α and β are fitting coefficients and V the filtrate volume. Because R_m is a function of the filtrate volume, and therefore dependent on time, its impact must be included in the integration of Equation (3.64). This is described in Zhao *et al.* (1998) with added detail in Bache and Zhao (2001). The semi-empirical approach described in Zhao *et al.* (1998) allows the component resistances (*i.e.* R_c and R_m) to be disentangled.

Making use of this framework, Zhao and Bache (2001) examined the interaction between the supracolloids and polymer residual in terms of their impact on the effective membrane resistance. This led to the conceptual model illustrated in Figure 3.42. For the particular system examined, it was found that there was no significant blinding when either residuals or supracolloids were absent. (cases (b) and (c)). Significant blinding (shown by increases in the membrane resistance) only occurred when both polymer residuals and supracolloids were present (case (d)). This was attributed to an interaction between the supracolloids and the residuals – leading to the formation of larger clusters which were more easily trapped. Trapping could be reduced by using a coarser filter (case (e)).

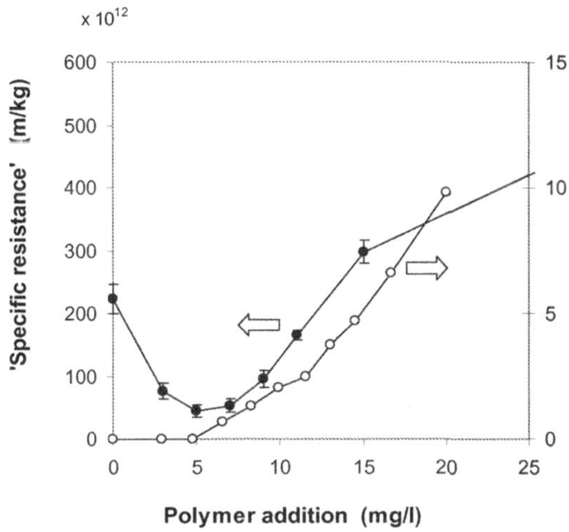

Figure 3.41. Shows the apparent 'specific resistance' of an alumino-humic sludge in response to polymer dose and the measured polymer residual. Note the emergence of residual polymer at the point of minimum resistance (Zhao and Bache, 2002).

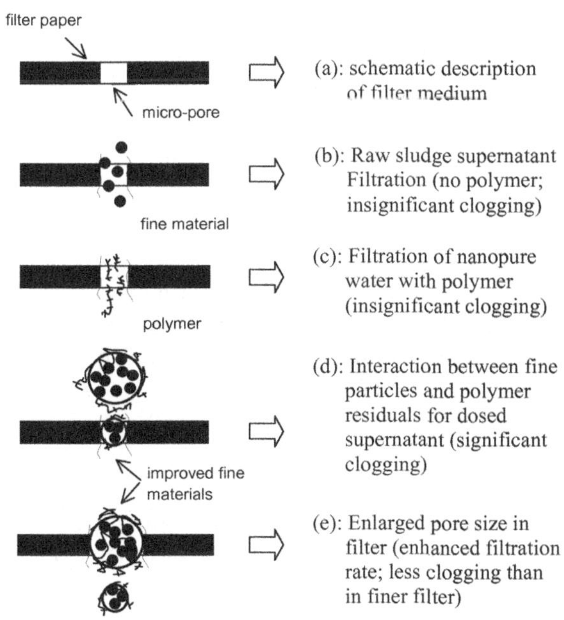

Figure 3.42. Conceptual view of clogging/blinding mechanism (Zhao and Bache, 2002).

3.3.7.10 Expression

Expression differs from filtration in that liquid is expelled from the cake or slurry by a reduction in its volume (see Figure 3.32). The extent of expression can be followed by monitoring the pressure ratio $P_L/\Delta P$ with P_L as the liquid pressure and ΔP the applied pressure difference across the cake (Novak *et al.*, 1999). Alternatively, this is often specified by the consolidation ratio:

$$U_c = \frac{L_o - L}{L_o - L_\infty} \tag{3.67}$$

in which L_o, L and L_∞ refer to the initial cake thickness, cake thickness at consolidation time θ_c and cake thickness when $\theta_c \to \infty$. In physical terms, U_c expresses the liquid mass released as a fraction of the total mass capable of release at the applied pressure.

Consolidation is considered to take place in two parts. Using the definitions from Wakeman (1981): "Primary consolidation refers to an instantaneous compression which occurs simultaneously with the increase in a local solids compressive pressure. Secondary consolidation refers to a delayed compression, which continues for a long time under a more or less constant particle pressure (sometimes referred to as the creep effect)". Shirato *et al.* (1970) adapted a Terzaghi-Voigt rheological model to show:

$$U_c = \underbrace{(1-B)\left\{1-\exp\left(-\frac{\pi^2 i^2 C_e}{4\omega_o^2}\theta_c\right)\right\}}_{\text{primary}} + \underbrace{B\{1-\exp(-\eta\theta_c)\}}_{\text{secondary}} \tag{3.68}$$

in which B is the fraction of moisture removed during secondary consolidation, i the number of drainage surfaces, C_e a modified consolidation coefficient reflecting, ω_o is the specific volume of a wet cake and η, a creep coefficient. When $\theta_c \to \infty$, Equation (3.68) reduces to:

$$U_c = 1 - B\exp(-\eta\theta_c) \tag{3.69}$$

A more recent development has been the evolution of a new class of macroscopic rheological models *e.g.* Landman and White (1994) stemming from Buscall and White (1987). These apply to expression as well as to other phases of dewatering. The framework is built round the concept of compressive yield stress to represent the network strength and a resistance term linked to the permeability.

At the microscopic level, studies such as Antelmi *et al.* (2001) and Madeline *et al.* (2006) provide detail of the mode of collapse during expression. This was achieved using small angle neutron scattering (Section 3.1.3.3). Under increasing pressure, small displacements lead to collapse of large pores in the first instance, then progressing through smaller pores. This behaviour appears to be uniquely determined by the strength of the interparticle bonds and geometric characteristics. One suspects that there are major parallels between these properties and those controlling floc strength.

Figure 3.43 illustrates the approach to the equilibrium state at two reference pressures for an alumino-humic sludge. Noting that the original sludge samples were 1 cm in depth (*i.e.* 'thin' in terms of typical cake thicknesses encountered at full scale), the trends shows that it takes several days to reach the equilibrium state in this type of sludge. Beyond the initial phase of dewatering (filtration), the trends display an 'exponential-like' response to time as indicated by Equation 3.69. At these longer timescales (*e.g.* beyond 4 days), one should be mindful of changing nature of the hydroxides (François, 1987) which will affect both permeability and consolidation behaviour (see also Section 3.2.7.2). The correspondence between the equilibrium state and the applied pressure is shown in Figure 3.44. In Figure 3.44, it is evident that the equilibrium moisture content reduces with increasing pressure (≡ compressive yield stress). The *extent of dewatering* shown by the equilibrium moisture content is unaffected by the polymer (NB the dose shown was optimal for the rate of release in the earlier stages of dewatering). The dewatering characteristics of the sludge and its precipitate fraction are similar – perhaps indicating that the precipitate dominates this phase of dewatering. However, Figure 3.43 shows that when the coagulant products were stripped out by acidifying the samples (*i.e.* leaving predominantly humic materials), the resulting moisture content is similar to the precipitate at the same pressure (casting doubt on the interpretation given above). This aspect is discussed further in Bache and Papavasilopoulos (2003), but remains as an imponderable.

Using a clay suspension coagulated with alum, Chang *et al.* (1997) provided detail of the response of secondary consolidation to coagulation conditions. As part of their analysis, Equation (3.69) was fitted to moisture release data with B and η treated as fitting coefficients. Their values were found to be dependent on dose and pH. By presenting data in the form a parameter-dose-pH plot, the combination of dose and pH, which were optimal for expression, could be identified. This was roughly coincident with a minimum in the dose-pH trend of a zero zeta potential - pointing yet again to the links between the long-term dewatering behaviour and the electrokinetic state (Sections 3.3.7.3 and 3.3.7.8).

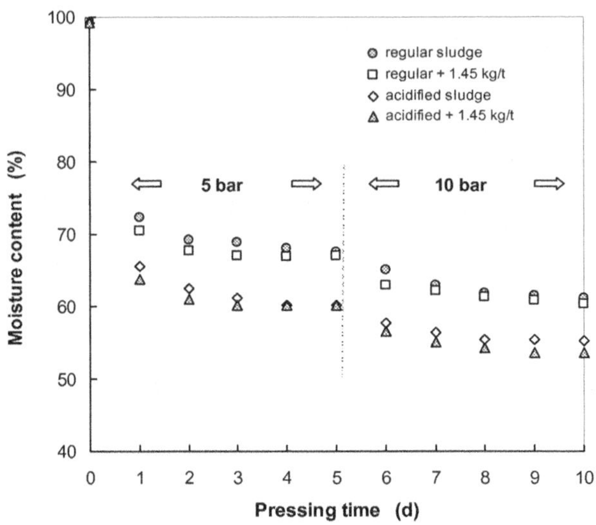

Figure 3.43. Dependence of moisture content on time and composition at two reference pressures; the value 1.45 kg/t refers to the optimum polymer dose (Bache and Papavasilopoulos, 2003).

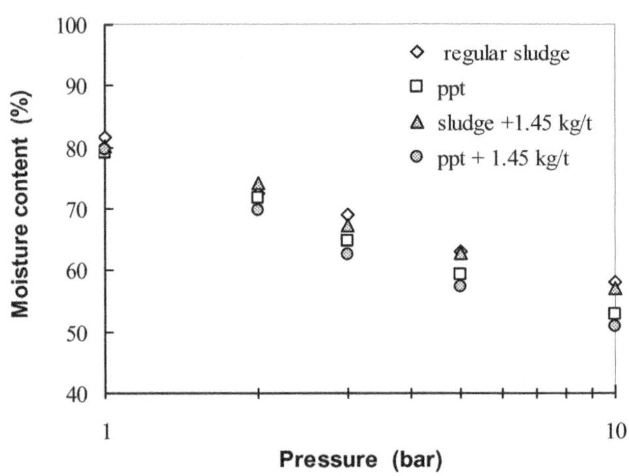

Figure 3.44. Equilibrium moisture content (MC) as a function of applied pressure for unconditioned (raw) and conditioned alumino-humic sludge (11,160 mg/l) and unconditioned and conditioned aluminium hydroxide precipitate (ppt; 4450 mg/l), this matching the coagulant fraction within the sludge; the value 1.45 kg/t refers to the optimum polymer dose (Bache and Papavasilopoulos, 2003).

3.3.7.11 *Key points*

When using Al and Fe salts as coagulants, many aspects of dewatering (including water content, polymer demand and filtration characteristics) can be linked to the hydroxide presence within the sludge. As the hydroxide fraction increases, the controls on dewatering become dominated by small scale behaviour, this being associated with the coagulant products.

Table 3.30. Examples of practical issues in sludge thickening and dewatering.

Symptom	Causes/Remedies
Thickening	
'Boiling' interface	Inequality of flow distribution, imbalance of flow geometry, excessive flow rate or blanket not high enough
Distinct interface	Should be able to see/detect distinct surface of blanket
Diffuse interface	May indicate that "maximum flux condition" is being exceeded *i.e.* excessive feed flow rate
High blanket level	Need to draw off solids } Improve monitoring/
Low blanket level	Suspend withdrawal of solids } control of blanket level
Settled turbidity high even though interface distinct	Perhaps blanket level too high Review coagulation chemistry; review age and handling of sludge before thickening; review polymer dose and application
Low thickened sludge concentration	Perhaps blanket too low/sludge removed too fast
Cake Filtration	
High pressure loss rate ~ shorter than expected filter runs	Improve permeability of sludge cake by increasing polymer dose and/or shear Blocking of filter fabric possibly due to polymer overdosing or poor dispersion.
Lower than expected cake solids concentration	Reduce rate of flow to filter press/drum
Poor quality filtrate	Check for optimum polymer dose/application
Centrifuging	
Lower than expected cake solids concentration	Improve strength of sludge by increasing polymer dose and/or dispersion of polymer. Reduce flow through centrifuge
Poor quality centrate	Check for optimum polymer dose/application
Optimisation of Polymer Dose	Ensure rapid and effective dispersion (but not excessive shear) Adjust headloss/shear for polymer dispersion for extremes of flow rate Adjust polymer dose to variation (> 10-20%) in solids concentration or link to coagulant metal-ion concentration Check polymer dose optimisation following major changes to raw water quality and coagulation chemistry

It is well known that polymers play a critical role in dewatering, with a major impact on the rate of release, but little impact on the extent of release. In coagulant sludge, the impact of polymers on the rate of release is most likely to emanate from the polymer-hydroxide interaction, on the basis that it is the permeability of the hydroxides which tends to be rate limiting. Apart from pattern of adsorption, the precise nature of the polymer-hydroxide interaction (insofar as it affects permeability) is unclear.

The polymer dose required for dewatering is governed by adsorption - particularly on the hydroxides. For cases in which polymer adsorption is dominated by the polymer-hydroxide interaction, it implies that the polymer dose is scaled by the coagulant dose [but compounded by factors such as shear]. When the adsorption capacity is exceeded, polymer residuals will be present.

Excess dosing of polymers and primary coagulants should be avoided in terms of dewaterability and operational problems such as filter blinding.

Consolidation phenomena are closely linked to the electrokinetic state of materials present, these affecting the yield stress.

Some practical issues are addressed in Table 3.30. Clearly this does not do justice to the wider array of problems that occur in practice. There may be need to address design and operation factors such reported in Bache (2000) or to implement procedures for minimising sludge production as discussed in Gregory and Dillon (2002). Overall, it demands a very good understanding of the issues involved.

3.3.8 Overview

- There is a need to secure far more information about the state of the solids phase within treatment processes; this should address size distributions in the first instance, as well as information about density and strength or rheological character.
- There is an acute lack of information about floc character in the approximate size range 1 – 50 μm, a range that is likely to play a critical role in separation processes.
- The example cited in Figure 3.31 emphasises the need to gain information about floc strength as well as size, in terms of their impact on treatability. Experience suggests that floc strength is associated with the base-line treatment rate (Table 3.14).
- Effective coagulation is an essential prerequisite to successful separation.
- Accurate dosing of coagulants and polymers is vital. Temptations to overdose should be avoided.

- In dewatering, there is a need to gain more information about interactions on the scales which control dewatering. Guidance to the dominating scale is provided by the cake permeability.

3.4 REFERENCES

Adin, A. and Rebhun, M. (1977) A model to predict concentration and lead-loss profiles in filtration. *J. AWWA.*, 69(8), 444 - 453

Al Farisi, A-L. (1988*) Effect of Bentonite Addition on Floc-Size Density Relationships in Coloured Water.* MSc. Thesis, University of Strathclyde, UK.

Arnold-Smith, A. K., Christie, R. M. and Jolicœur, C. (1992) Polyaluminium silicate sulphate – a new coagulant for potable and wastewater treatment. In*: Chemical Water and Wastewater Treatment II* (Eds: Klute, R. and Hahn, H.H.), 203-219, Springer-Verlag, Berlin.

Amirtharajah, A. and Mills, K. M. (1982) Rapid-mix designs for mechanisms of alum coagulation. *J. AWWA*, 74(4), 210-216.

Amirtharajah A & O'Melia CR (1990) Coagulation processes: destabilization, mixing and flocculation. In: *Water Quality & Treatment*, AWWA 4th ed., McGraw Hill, 269-365.

Amirtharajah, A. and Tambo, N. (1991) Mixing in water treatment. In: *Mixing in Coagulation and Flocculation* (Eds Amirtharajah, A., Clark, M.M. and Trussell, R.R.) AWWA Res. Foundation, Denver Co, 3-34.

Ammary, B.Y. and Cleasby, J. L., (1995) Flocculation kinetics using cationic polymers in combination with a ferric salt. *Annual Conf. Proc.*, AWWA, Paper No. ACE95091, 953-996.

Antelmi, D., Cabane, B., Meireles, M. and Almar, P. (2001) Cake collapse in pressure filtration. *Langmuir*, 17, 7134-7144.

Aubert, C. and Cannell, D. S. (1986) Restructuring of colloidal silica aggregates. *Phys. Rev. Lett.*, 56(7), 738-741.

AWWA (1992) *Operational Control of Coagulation and Filtration Processes*. AWWA Manual M37. 1st Edn., AWWA, Denver, Co.

Bache, D.H. (1984) Prediction of the bulk deposition velocity and concentration profiles within plant canopies. *Atmos. Environ.*, 18(11), 2517-2519.

Bache, D. H. (2002) Polymers in alum sludge dewatering: development and issues. In: *Chemical Water and Wastewater Treatment VII*, (Hahn, H.H., Hoffmann, E. and Ødegaard, H., Eds), IWA Publishing, London, 287-296.

Bache, D.H. (2004) Floc rupture and turbulence: a framework for analysis. *Chem. Eng. Sci.*, 59. 2521-2534.

Bache, D. H., Ali. A., Rasool, R., McGilligan, J.F. and Johnson, C. (1996) Temperature and coagulation control in the sweep floc domain. *J. Water SRT – Aqua*, 45(4), 195-202.

Bache, D. H. and Hossain, M. D. (1991) Optimum coagulation conditions of a coloured water in terms of floc properties. *J. Water SRT-Aqua*, 40(3), 170-178.

Bache, D.H., Johnson, C., Papavasilopoulos, E., Rasool, E. and McGilligan, J. F. (1999a) Sweep coagulation: structures, mechanisms and practice. *J. Water SRT- Aqua*, 48(5), 201-210.

Bache, D. H. and Papavasilopoulos, E. N. (2000) Viscous behaviour of sludge centrate in response to polymer conditioning. *Water Res.*, 34, 354-358.

Bache, D.H. and Papavasilopoulos (2003) Dewatering of alumino-humic sludge: impacts of hydroxide. *Water Res.*, 37, 3289-3298.

Bache, D.H. and Rasool, E. (1996) Measurement of the rate of energy dissipation around an oscillating grid by an energy balance approach. *Chem. Eng. J.*, 66, 105-115.

Bache, D. H. and Rasool, E. (2001a) Characteristics of alumino-humic flocs in relation to DAF performance. *Water Sci. Tech.*, 43(8), 203-208.

Bache, D.H. and Rasool, E,. (2001b) Characteristics of turbulence in a multigrid mixer. *Chem. Eng. J.*, 83, 67-78.

Bache, D.H. and Rasool, E. (2006) Characterisation, scaling and analysis of steady state floc size distribution using a mass balance approach. *J. Water SRT-Aqua*, 55(5), 335-356.

Bache, D.H., Rasool, E., Ali, A. and McGilligan, J. F. (1995) Floc character: measurement and role in optimum dosing. *J. Water SRT- Aqua*, 44(2), 83-92.

Bache, D.H., Rasool, E., Moffat, D. and Johnson, M. (2000) Floc characterisation at full scale: the case of humic waters. In: *Chemical Water and Wastewater Treatment VI*, (Hahn, H.H., Hoffmann, E. and Ødegaard, H., Eds), Springer, Heidelberg, 66-76.

Bache, D.H., Rasool, E., Moffat, D. and McGilligan, F. J. (1999b) On the strength and character of alumino-humic flocs. *Water Sci. Tech.*, 40(9), 81-88.

Bache , D. H. and Zhao, Y. Q. (2001) Optimising polymer use in alum sludge conditioning: an *ad hoc* test. *J. Water SRT-Aqua*, 50(1), 29-38.

Bai, R. and Tien, C. (1997) Particle detachment in deep bed filtration. *J. Colloid Interface Sci.*, 186, 307-317.

Baldyga, J. and Bourne, J. R. (1995) Interpretations of turbulent mixing using fractals and multifractals. *Chem. Eng. Sci.*, 50(3), 381-400.

Baylis, J. R. (1939) Silicates as aids to coagulation. *J. AWWA*, 29, 1355-1396.

Besra, L., Sengupta, D. K., Roy, S. K. and Ay, P. (2002) Flocculation and dewatering of kaolin suspensions in the presence of polyacrylamide and surfactants. *Int. J. Miner. Process.*, 66, 203-232.

Bernhardt, H., Lüsse, B. and Hoyer, O. (1986) The addition of calcium to reduce the impairment of flocculation by algogenic organic matter. *Zeitschrift fur wasser und abwasser forschung – J. Water and Wastewater Res.*,19(6), 219-228.

Blaser, S. (2000a) Break-up of flocs in contraction and swirling flows. *Colloids and Surfaces A,* 166, 215-223.

Blaser, S. (2000b) Flocs in shear and strain flows. *J. Colloid Interface Sci.*, 225, 273-284.

Bolto, B., Dixon, D., Eldridge, R. and King, S. (2001) Cationic polymer and clay or metal oxide combinations for natural organic matter removal. *Water Res.*, 35(11), 2669-2676.

Bottero, J. Y. and Fiessinger, F. (1989) Aluminium chemistry in aqueous solution. *Nordic Pulp and Paper Research*, 2, 81- 89.

Bottero, J. Y., Tchoubar, D., Arnaud, M. and Quienne, P (1991) Partial hydrolysis of ferric nitrate salt. Structural investigation by dynamic light scattering and small-angle X-ray scattering. *Langmuir*, 7, 1365-1369.

Bratby, J. R. (1981) Optimising coagulants and flocculant aids for settling. *J. AWWA*, 73(6), 312-318 (and *Erratum* 73(11), p 38).

Buscall, R. and White, L. R. (1987) The consolidation of concentrated suspensions. Part 1.- The theory of sedimentation. *J. Chem. Soc., Faraday Trans. 1*, 83, 873-891.

Bushell, G. C., Yan, Y. D., Woodfield, D., Raper, J. and Amal, R. (2002) On techniques for the measurement of the mass fractal dimension of aggregates. *Adv. Colloid Interface Sci.*, 95, 1-50.

Cathalifaud, G., Wais Mossa, M. T. and Mazet, M. (1993) Preformed ferric hydroxide flocs as adsorbents of humic substances. *Water Sci. Tech.*, 27(11), 55-60.

Camp, T. R., Root, D. A. and Bhoota, B. V. (1940) Effect of temperature on rate of floc formation . *J. AWWA*, 32(11), 1913-1927.

Chakraborti, R.K., Atkison, J.F. and van Benschoten, J.E. (2000) Characterization of alum floc by image analysis. *Environ. Sci. Tech.*, 34, 3969-3976.

Chakraborti, R.K., Gardner, K.H., Atkison, J.F. and van Benschoten, J.E. (2003) Changes in fractal dimensions during aggregation. *Water Res.*, 37(4), 873-883.

Chang, I. L., Chu, C. P., Lee, D. J. and Huang, C. (1997) Expression dewatering of alum-coagulated clay slurries. *Environ. Sci. Technol.*, 31(5), 1313-1319.

Chen, G. W., Hung, W. T., Chang, I. L., Lee, S. F. and Lee, D. J. (1997) Continuous classification of moisture content of waste activated sludges. *J. Environ. Eng.*, 123(3), 253-258.

Chen, S. H. and Teixeira, J. (1986) Structure and fractal dimension of protein-detergent complexes. *Phys. Rev. Lett.*, 57(20), 2583-2586.

Cheremisinoff, P. N. and Ellerbusch, F. (1978) *Carbon Adsorption Handbook.* (Eds. Cheremisinoff, P. N. and Ellerbusch, F.) Ann Arbour Science, Ann Arbour, Mi.

Clark, M. M. and Fiessinger, F. (1991) Mixing and scale-up. In: *Mixing in Coagulation and Flocculation* (Eds Amirtharajah, A., Clark, M.M. and Trussell, R.R.) AWWA Res. Foundation, Denver Co, 282-306.

Clark, M. M. and Flora, J. R. V. (1991) Floc restructuring in varied turbulent mixing. *J. Colloid Interface Sci.*, 147(2), 407-421.

Cleasby, J. L. (1990) Filtration. In: *Water Quality and Treatment* (Pontius, F. W., Ed), AWWA 4th ed., McGaw-Hill, 455-560.

Coackley, P. and Jones, B.R. S. (1956) Vacuum sludge filtration: I. Interpretation of results by the concept of specific resistance. *Sew. & Ind. Wastes*, 28(8), 963-975

Cornwell, D. A. and Bishop, M. M. (1983) Determining velocity-gradients in laboratory and full-scale systems. *J. AWWA.*, 75(9), 470-475.

CRC(1980) *Handbook of Chemistry and Physics*, 61st edition, CRC Press, Boca Raton, Florida.

Dai, X and Boll, J. (2006) Settling velocity of Cryptosporidium parvum and Giardia lamblia. *Water Res.*, 40, 1321-1325.

Dauthuille, P. (1992) The DENSADEG – a new high performance settling tank. In: *Chemical Water and Wastewater Treatment II.* (Eds. Klute, R. and Hahn, H. H.), Springer-Verlag, Berlin., 135-150.

de Boer, G. B. J., Hoedmakers, G. F. M. and Thoenes, D (1989) Coagulation in turbulent flow: Part I. *Chem. Eng. Res. Des.*, 67, 301-315.

de Dianous, F., Pujol, E. and Druoton, J. C. (1990) Industrial application of weighted flocculation: development of the Actiflo Clarification Process. In: *Chemical Water and Wastewater Treatment.* (Eds. Hahn, H. H. and Klute, R.), Springer-Verlag, Heidelberg.

Degrémont, S. A. (1991) *Water Treatment Handbook.* 6th ed., Degrémont, Paris.

Dempsey, B.A. (1984), Removal of naturally occurring compounds by coagulation and sedimentation. *CRC Critical Reviews in environmental Control,* 14(4), 311-331.

Dempsey, B. A. (1987) Chemistry of coagulants. Proceedings of seminar on influence of coagulation on the selection, operation and performance of water treatment facilities. AWWA Denver, Co., 31-47.

Dentel, S. K. (1991) Coagulation control in water treatment. *Critical Reviews in Environmental Control*, 21(1), 41-135.

Dentel, S. K. and Abu-Orf, M. M. (1995) Laboratory and full scale studies of liquid stream viscosity and streaming current for characterization and monitoring of dewaterability. *Water Res.*, 29(12), 2663-2672.

Dentel, S. K., Gucciardi, B. M., Bober, T. A., Shetty, P. V. and Resta, J. J. (1989) *Procedures Manual for Polymer Selection in Water Treatment Plants.* AWWA Research Foundation, Denver, Co.

Dentel, S. K., Resta, J. J., Shetty, P. V. and Bober, T. A. (1988) Selecting coagulant, filtration and sludge conditioning aids. *J. AWWA*, 80(1), 72-84.

Desjardins, C, Koudjonou, B. and Desjardins, R. (2002) Laboratory study of ballasted flocculation, *Water Res.*, 36(16), 744-754.

Dillon, G. (2004) *Optimised Waterworks Sludge Thickening,* Report P6389/1, WRc plc Swindon, UK. (Not in public domain).

Divakaran, R. and Pillai. V.N.S. (2002) Flocculation of river silt using chitosan, *Water Res*. 36(9), 2414-2418.

Dolejs, P. (1984) Interaction of temperature, alkalinity and alum dose by coagulation of humic water. In: Pawlowski, L., Verdier, A.J. and Lacy W. J. eds. *Chemistry for Protection of the Environment*. Amsterdam, Elsevier, 169-178.

Duan, J. and Gregory, J. (1996) Influence of soluble silica on coagulation by aluminium sulphate. *Colloids and Surfaces A*, 107, 309-319.

Duan, J and Gregory, J. (1998) The influence of silicic acid on aluminium hydroxide precipitation and flocculation by aluminium salts. *J. Inorg. Biochem.*, 69, 193-201.

Duan, J. and Gregory, J. (2003) Coagulation of hydrolysing metal salts. *Adv. Colloid Interface Sci.*,100-102, 475-502.

Dulin, B. E. and Knocke, W. R. (1989) The impact of incorporated organic matter on the dewatering characteristics of aluminium hydroxide sludges. *J. AWWA.*, 81(5), 74-79.

Edzwald, J. K. (1993) Coagulation in drinking water treatment: particles, organics and coagulants. *Water Sci. Tech.*, 27(11), 21-35.

Edzwald, J. K. (1995) Principles and applications of dissolved air flotation. *Water Sci. Tech.*, 31(3-4), 1-23.

Edzwald, J. K., Pernitsky, D. J. and Parmenter, W. L. (2000) Polyaluminium coagulants for drinking water treatment: chemistry and selection. In: *Chemical Water and Wastewater Treatment VI* (Eds: Hahn, H.H., Hoffmann, E and Ødegaard, H.), pp 3-14, Springer-Verlag, Berlin.

Edzwald, J. K. and Tobiason, J. E. (1998) Enhanced versus optimised multiple objective coagulation. In: *Chemical Water and Wastewater Treatment V* (Eds: Hahn, H.H., Hoffmann, E and Ǿdegaard, H.), pp 173-187, Springer-Verlag, Berlin.

Eikebrokk, R. (1996) Removal of humic substances by coagulation. In: *Chemical Water and Wastewater Treatment IV* (Eds: Hahn, H.H., Hoffmann, E and Ødegaard, H.), pp 113-124, Springer-Verlag, Berlin.

Elimelech, M, Gregory, J., Jia, X. and Williams, R.A. (1995) *Particle Deposition and Aggregation*. Butterworth-Heinemann, Oxford.

Fettig, J., Miethe, M. and Kassebaum, F. (1996) Coagulation and precipitation by an alkaline aluminium coagulant. In: *Chemical Water and Wastewater Treatment IV* (Eds: Hahn, H.H., Hoffmann, E and Ødegaard, H.), pp 107-117, Springer-Verlag, Berlin.

Firth, B. A. (1976) Flow properties of coagulated colloidal suspensions II. Experimental properties of the flow curve parameters. *J. Colloid Interf. Sci.*, 57(2), 257-265.

François, R. J. (1987) Ageing of aluminium hydroxide flocs. *Water Res.*, 21(5), 523-531.

François, R. J. and Van Haute, A. A. (1984) The role of rapid mixing on a flocculation process. *Water Sci. Tech.*, 17 (6/7), 1091-1101.

Fukushi K., Tambo N. and Matsui Y. (1995) A kinetic model for dissolved air flotation in water and wastewater treatment. *Water Sci. Tech.*, 31 (3-4), 37-47.

Fukushi, K., Matsui, Y. and Tambo, N. (1998) Dissolved air flotation: experiments and kinetic analysis. *J. Water SRT-Aqua*, 47(2), 76-86.

Gao, B., Yue, Q., Zhao, H. and Song, Y. (2000) Properties and evaluation of polyferric-silicate-sulphate (PFSS) as a coagulant for water treatment. In: *Chemical Water and Wastewater Treatment VI* (Eds: Hahn, H.H., Hoffmann, E and Ødegaard, H.), pp 15-22, Springer-Verlag, Berlin.

Gibbs, R. J. (1982a) Floc stability during Coulter counter size analysis. *J. Sediment. Petrol.*, 52(2), 657-660.

Gibbs, R. J. (1982b) Floc breakage during Hiac light blocking analysis. *Environ. Sci. Technol.*, 16, 298-299.

Gillberg, L. (1994) Influence of basicity of polyaluminium chlorides when cleaning municipal wastewater. In: *Chemical Water and Wastewater Treatment III* (Eds: Klute, R. and Hahn, H.H.), pp 39-56, Springer-Verlag, Berlin.

Glasgow, L. A. and Hsu, J. P. (1982) An experimental study of floc strength. *AIChE J.*, 28(5), 779-785.

Glasgow, L. A., Kim, Y. H. and Hsu, J. P. (1986) Characterisation of turbulence-induced aggregate breakage. In: *Flocculation, Sedimentation and Consolidation.* (Eds Moudgil, B. M . and Somasundaran, P.), Proc. Engineering Foundation Conf., AIChE, New York, 191-204.

Glatter, O (2004) The inverse-scattering problem in small-angle scattering. In: *Neutrons, X-rays and Light: Scattering Methods Applied to Soft Condensed Matter*. (Eds Lindner, P and Zemb, Th). Second impression. Elsevier, Amsterdam.

Gregor, J.E., Nokes, C.J. and Fenton, E. (1997) Optimising natural organic matter removal from low turbidity waters by controlled pH adjustment of aluminium coagulation. *Water Res.*, 31(12), 2949-2958.

Gregory, J. (2004) Monitoring floc formation and breakage. *Water Sci. Tech.*, 50(12), 163-170.

Gregory, J. and Nelson, D. W. (1986) Monitoring of aggregates in flowing suspensions. *Colloids and Surfaces*, 18(2-4), 175-188.

Gregory, J. and Yukselen, M. A. (2002) Break-up and re-formation of flocs formed by hydrolysing coagulants and polymeric flocculants. In: *Chemical Water and Wastewater Treatment VII* (Eds: Hahn, H.H., Hoffmann, E and Ødegaard, H.), pp 29-38, IWA Publishing, London.

Gregory, R. (1977) *Floc blanket clarification*. M. Phil. Thesis, London University.

Gregory R. (1979) *Floc blanket clarification,* Report TR111, WRc, Swindon, UK
Gregory, R. (1987) *Eccup Process Performance Evaluation: Mixing and Flocculation Velocity Gradient Audit.* Report 659S, WRc plc, Swindon, UK.
Gregory R (1991) *Controlling coagulant residuals: Direct filtration – operation and performance,* Rep. No. UM1273, WRc, Swindon, UK.
Gregory, R, Connor, K. and Hullance, J. (1996a) *Better Use of Coagulants and Polyelectrolytes in Water Treatment*, Rep. No. PT 2025, WRc, Swindon, UK.
Gregory, R. and Dillon, D (2002) Minimising sludge production at water-treatment plants. *J. Chart. Inst. Water Environ. Manag.*, 16, 174-179.
Gregory, R., Head, R. J. M. and Graham, N. J. D. (1996b) The relevance of blanket solids concentration in understanding the performance of floc blanket clarifiers in water treatment. In: *Chemical Water and Wastewater Treatment VI*, (Hahn, H.H., Hoffmann, E. and Odegaard, H., Eds), Springer, Heidelberg, 17-29.
Gregory R., Maloney, R. J. and Stockley, M. (1988) Water treatment using magnetite: A study of a Sirofloc pilot plant, *J. Inst. Water Environ. Mngmt*, 2(5), 532-544.
Gregory, R. and Miller, R. J. (1994) *Selection and Use of Coagulants and Polyelectrolyte Flocculants.* Rep. No. PT1033, WRc Swindon, UK.
Gregory, R. Zabel, T.F. and Edzwald, J.K. (1999) Sedimentation and flotation. Chapt 7 in: *Water Quality and Treatment: A Handbook of Community Water Supplies.* (Letterman, R. D., Ed) AWWA 5th ed., McGraw-Hill, New York
Hall, T. (1997) *Water Treatment Processes and Practices.* Second Edition. WRc, Swindon, UK.
Han, M. Y. (2002) Modeling of DAF: the effect of particle and bubble characteristics. *J. Water SRT-Aqua*, 51(1), 27-34.
Hanson, A. T. and Cleasby, J. L. (1990) The effects of temperature on turbulent flocculation: Fluid dynamics and chemistry. *J. AWWA*, 82(11), 56-73.
Hayashi, N., Iwata, M., Kato, I., Takemura, K., Murase, T. and Shirato, M. (1990) Effect of residual polymer flocculant on the CST test for determining the optimum dosage. *Int. Chem. Eng.*, 30(4), 691-694.
Head, R., Hart, J. and Graham, N. (1997) Simulating the effect of blanket characteristics on the floc blanket clarification process. *Water Sci. Tech.*, 36(4), 77-84.
Hill, R. and Power, G. (1956) Extremum principles for slow viscous flow and the approximate calculation of drag. *Quart. J. Mech. Appl. Math.*, 9, 313-319.
Hossain, M.D. and Bache, D.H. (1991) Composition of alum flocs derived from a coloured low-turbidity water. *J. Water SRT-Aqua*, 40, 298-303.
Hudson, H.E. (1981) *Water Clarification Processes: Practical Design and Evaluation.* Van Nostrand Reinhold, New York.
Ives, K. J. (1968) Theory of operation of sludge blanket clarifier. *Proc. Inst. Civ. Eng.*, 39, 243-260.
Ives, K.J. (1975) Mathematical models of deep bed filtration. In: *The Scientific Basis of Filtration* (Ives, K. J., Ed), Noordhoff International Publishing, ISBN 90 286 0523 1, 203-224.
Jeffcoat, W. B. and Singley, J. E. (1975) The effect of alum concentration and chemical-addition times on coagulation. *J. AWWA*, 67, 177-181.
Jefferson, B., Jarvis, P. and Parsons, S.A. (2004) The effect of coagulant type on natural organic matter floc structure and strength. In: *Chemical Water and Wastewater*

Treatment VIII (Eds: Hahn, H.H., Hoffmann, E and Ødegaard, H.), pp 151-159, IWA Publishing, London..

Jiang, J. Q. and Graham, N.J. D. (1998) Evaluation of poly-alumino-iron sulphate (PAFS) as a coagulant for water treatment. In: *Chemical Water and Wastewater Treatment V* (Eds: Hahn, H.H., Hoffmann, E and Ødegaard, H.), pp 15-24, Springer-Verlag, Berlin.

Johnson, C. (1998) *Chemical factors in floc formation and structure*. PhD thesis, University of Strathclyde, Glasgow, UK.

Johnson, S. B., Dixon, D. R. and Scales, P. J. (1999) The electrokinetic and shear yield stress properties of kaolinite in the presence of aluminium ions, *Colloids Surf. A*, 146, 281-291.

Jung, S. J., Amal, R. and Raper, J. A. (1996) Monitoring effects of shearing on floc structure using small-angle light scattering. *Powder Tech.*, 88, 51-54.

Kam, S-K. and Gregory, J. (2001) The interaction of humic substances with cationic polyelectrolytes. *Water Res.*, 35(15), 3557-3566.

Kang, L. and Cleasby, J. L. (1995) Temperature effects on flocculation kinetics using Fe(III) coagulant. *J. Environ Engng ASCE*, 121(12) 893-901.

Karr, P. R. and Keinath, T. M. (1978) Influence of particle size on sludge dewaterability. *J. Water Poll. Control Fed.*, 8, 1911-1929.

Knocke, W. R., Hamon, J. R. and Dulin, B. E. (1987) Effects of coagulation on sludge thickening and dewatering. *J. AWWA.*, 79(6), 89-98.

Knocke, W. R. and Wakeland, D. L. (1983) Fundamental characteristics of water treatment plant sludges. *J. AWWA.*, 75(10), 516-523.

Kawamura, S. (1973) Coagulation considerations. *J. AWWA*, 65, 413-423.

Kawamura, S. (1996) Optimisation of basic water treatment processes – design and operation; coagulation and flocculation. *J. Water SRT- Aqua.*, 45(1), 35-47.

Keenan, H., Papavasilopoulos, E. N. and Bache, D. H. (1998) On the measurement of polymer residuals following the conditioning of an alum sludge. *Water Res.*, 32(10), 3173-3176.

Kim, A. Y. and Berg, J. C. (1999) Fractal aggregation: scaling of the fractal dimension with stability ratio. *Langmuir*, 16(5), 2101-2104.

Kiuru, H. J. (2001) Developments of dissolved air flotation technology from the first generation to the newest (third) one (DAF in turbulent flow conditions). *Water Sci. Tech.*, 43(8), 1-7.

Klute, R. and Amirtharajah, A. (1991) Particle destabilization and flocculation reactions in turbulent pipe flow. In: *Mixing in Coagulation and Flocculation* (Eds Amirtharajah, A., Clark, M.M. and Trussell, R.R.) AWWA Res. Foundation, Denver Co, 217-255.

Konno, H. (1993) Settling and coagulation of slender type diatoms. *Water Sci. Tech.*, 27(11), 339-346.

Landman, K.A. and White, L. R. (1994) Solid/liquid separation of flocculated suspensions. *Adv. Colloid Interface Sci.*, 51, 175-246.

Lawler, D. F. and Nason, J. A. (2004) The scientific basis of flocculator design. *Water Sci. Tech.*, 50(12),155-162.

Lawler, D. F. and Nason, J. A. (2006) Granular media filtration: old process, new thoughts. *Water Sci. Tech.*, 53(7), 1-7.

Lee, D. J., Chen, G. W., Liao, Y. C. and Hsieh, C. C. (1996) On the free-settling test for estimating activated sludge floc density. *Water Res.*, 30(3), 541-550.
Letterman, R.D., Quon, J.E. and Gemmell, R. S. (1973) Influence of rapid-mix parameters on flocculation. *J. AWWA*, 65(11), 716-722.
Letterman, R.D. and Sricharoenchaikit, P., (1982) Interaction hydrolized Al and polyelectrolyte coagulants, *Proc. ASCE.*, *J. Environ. Engrg. Div.*, 108(EE5), 883-899.
Li, D. H. and Ganczararczyk, J. (1989) Fractal approach to floc structure and the floccualtion process. *Proc. Annual Conf. Am. Water Works Assoc.*, Los Angeles, Calif., 1433-1442.
Lindqvist, N., Jokela, J. and Tuhkanen, T. (2004) Enhancement of coagulation using polyelectrolytes as coagulant or flocculant aids. In*: Chemical Water and Wastewater Treatment VIII* (Eds: Hahn, H.H., Hoffmann, E and Ødegaard, H.), pp 161-169, IWA Publishing, London.
Lockyear, C. F., Jackson, P. J. and Warden, J. H. (1983) *Polyelectrolyte User's Manual.* Rep. No. TR 184, WRc, Swindon, UK.
Logan, B. E. and Kilps, J. R. (1995) Fractal dimensions of aggregates formed in different fluid mechanical environments. *Water Res.*, 29(2), 443-453.
Madeline, J. B., Meireles, M., Botet, R. and Cabane, B. (2006) The role of interparticle forces in colloidal aggregates: local investigations and modeling of restructuring during filtration. *Water Sci. Tech.*, 53(7), 25-32.
Masschelein, W. J. (1992) *Unit Processes in Drinking Water Treatment.* Marcel Dekker, New York.
Matsui, Y., Yuasa, A., Furuya, Y. and Kamei, T. (1998) Dynamic analysis of coagulation with alum and PACl. *J. AWWA*, 90(10), 96-106.
Matsumoto, K., Suganuma, A. and Kunii, D. (1978) Effect of permeability on the settling velocity of an actual floc. *Chem. Eng. Sci.*, 33, 1554-1556.
Mazet, M., Angbo, L. and Serpaud, B. (1990) Adsorption of humic acids onto preformed aluminium hydroxide flocs. *Water Res.*, 24(12), 1509-1518.
Meakin, P. (1988) Fractal aggregates. *Adv. Colloid Interface Sci.*, 28(4), 249-331.
Moffet, J. W. (1968) The chemistry of high-rate water treatment. *J. AWWA*, 60, 1255-1270.
Montgomery, J. M.* (1985) *Water Treatment Principles and Design*, Wiley, New York. (* also classified as "J. M. Montgomery, Consulting Engineers").
Morris, J.K. and Knocke, W.R. (1984) Temperature effects on the use of metal-ion coagulants for water treatment, *J. AWWA.*, 76, 74-79.
Morrow, J.J. and Rausch, E.G. (1974) Colloid destabilisation with cationic polyelectrolytes as affected by velocity gradients, *J. AWWA*. 66(11), 646-653.
Mpofu, P., Addai-Mensah, J. and Ralston, J. (2003) Influence of hydrolyzable metal ions on the interfacial chemistry, particle interactions, and dewatering behavior of kaolinite dispersions. *J. Colloid Interface Sci.*, 261, 349-359.
Mutl, S., Knesl, B. and Polasek, P., (2002) Factors influencing direct separation of destabilised impurities in a fluidised layer of granular material. *J. Water SRT- Aqua*, 51(1), 15-26.
Ndabigengesere, A., Narasiah, K.S. and Talbot, B.G. (1995) Active agents and mechanism of coagulation of turbid waters using Moringa Oleifera. *Water. Res.* 29(2), 703-710.

Nickel, J. B. (1945) Fundamental technics for running jar tests. *J. AWWA*, 60, 1255-1270.

Novak, J. T., Agerbæk, M. L., Sorensen, B. L. and Hansen, J. Aa. (1999) Conditioning, filtering, and expressing waste activated sludge. *J. Environ. Eng.*, 125(9), 816-824.

Novak, J. T. and Bandak, N. (1997) The effect of shear on the dewatering of water treatment residuals. *J. AWWA.*, 88(11), 84-91.

Novak, J. T., Goodman, G. L., Pariroo, A. and Juang, J-C. (1988) The blinding of sludges during filtration. *J. Water Pollut. Contr. Fed.*, 60(2), 206-214.

Novak, J. T. and O'Brien, J. H. (1975) Polymer conditioning of chemical sludges. *J. Water Pollut. Contr. Fed.*, 47(10), 2397-2410.

O'Melia, C. R. (1985) Particles, pre-treatment, and performance in water filtration. *J. Env. Eng.*, 111(6), 774-890.

Olin, H. L. and Peterson, H. W. (1937) The use of bentonite as a coagulant in water treatment. *J. AWWA*, 29, 513-522.

Okamoto, Y., Nishikawa, M. and Hashimoto, K. (1981) Energy dissipation rate distribution in mixing vessels and its effects on liquid-liquid dispersion and solid-liquid mass transfer. *International Chem. Eng.*, 21(1), 88-94.

Packham, R. F. and Saunders, M. (1966) Laboratory studies of activated silica – 1: Effect of the coagulation of a clay with aluminium sulphate. *Proc. Soc. Water Treat. Exam.*, 15, 245-254.

Papavasilopoulos, E. N. and Bache, D. H. (1998) On the role of aluminium hydroxide in the conditioning of an alum sludge. *Water Sci. Tech.*, 38(2), 33-40.

Parazak, D.P., Burkhardt, C.W., McCarthy, K.J. and Stehlin, M.P. (1987) Hydrophobic flocculation. *J Colloid Interface Sci.*, 123(1), 59-72.

Patel, N. S. and Shah, A. N. (1978) Coagulant aids – a review. *J. Ind. Water Works Assoc.*, X(4), 356-362.

Pedersen, J. S. (2004) Modelling of small-angle scattering data from colloids and polymer systems. In: *Neutrons, X-rays and Light: Scattering Methods Applied to Soft Condensed Matter.* (Eds Lindner, P and Zemb, Th). Second impression. Elsevier, Amsterdam

Perry, J. H. (1950) *Chemical Engineers Handbook.* 3rd ed., McGraw-Hill, New York.

Probstein, R. F. (1994) *Physicochemical Hydrodynamics.* 2nd ed. Wiley-Interscience, New York.

Pruden, B. B. and Epstein, N. (1964) Stratification by size in particulate fluidisation and in hindered settling. *Chem Eng. Sci.*, 19, 696-700.

Pusey, P. N. (2004) Introduction to scattering experiments. In: *Neutrons, X-rays and Light: Scattering Methods Applied to Soft Condensed Matter.* (Eds Lindner, P and Zemb, Th). Second impression. Elsevier, Amsterdam

Rasmussen, F. B. (2001) Different approaches to the analysis of small angle scattering experiments on porous aluminium hydroxide. *Colloids Surf. A*, 187-188, 327-335.

Rebhun, M., Fuhrer, Z. and Adin, A. (1984) Contact flocculation-filtration of humic substances. *Water Res.*, 18(8), 963-970.

Richard, Y. and Capon, B. (1980) Sedimentation. In: *Developments in Water Treatment –1*, (Ed: Lewis, W. M.) Applied Science Publishers.

Selomulya, C., Bushell, G., Amal, A. and Waite, T.D. (2003) Understanding the role of restructuring in flocculation: the application of a population balance model. *Chem. Eng. Sci.*, 58, 327-338.

Shirato, M., Muase, T., Kato, H. and Fukaya, S. (1970) Fundamental analysis for expression under constant pressure. *Filt. Sep.*, 7(3), 277-282.

Sibony J. (1981) Clarification with microsand seeding – a state of the art. *Water Res.*, 15(11), 1281-1290.

Singley, J. E. (1972) Theory and mechanism of polyelectrolytes as coagulant aids. In: AWWA Seminar on *Polyelectrolytes Aids to Better Water Quality*, Chicago. American Water Works Assoc., New York., Paper 2.

Spicer, P. T., Pratsinis, S. E., Raper, J., Amal, R., Bushell, G. and Meesters, G. (1998) Effect of shear schedule on particle size, density and structure during flocculation in stirred tanks. *Powder Tech.*, 97, 26-34.

Standen, G., Insole, P. J., Shek, K. J. and Irwin, R. A. (1995) Optimisation of clarifier performance for pesticide removal. In: *Water and Waste Water Treatment – Towards Cleaner, Leaner Operation* (Proc. 3rd Int. Conf., Harrogate, England), BHR Group Conf. Series Pub. No. 17, ISBN 0852989776. Mechanical Engineering Publications, Ltd., London.

Stanley, S.J. and Smith, D.W. (1995) Measurement of turbulent flow in standard jar test apparatus. *J Environmental Engineering* 121(12), 902-909.

Su S.T., Wu R. M. and Lee, D.J. (2004) Blanket dynamics in upflow suspended bed, *Water Res.* 38(1), 89-96

Tambo, N and Hozumi, H. (1979) Physical aspect of flocculation process – II. Contact flocculation. *Water Res.*, 13, 441-448.

Tambo, N. and Wang, X. C. (1993) The mechanism of pellet flocculation in a fluidized-bed operation. *J. Water SRT-Aqua*, 42(2), 67-76.

Tambo, N. and Watanabe, Y. (1979) Physical characteristics of flocs-I. The floc density function and aluminium floc. *Water Res.*, 13, 409-419.

Tambo, N. and François, R. J. (1991) Mixing, breakup, and floc characteristics. In: *Mixing in Coagulation and Flocculation* (Eds Amirtharajah, A., Clark, M.M. and Trussell, R.R.) AWWA Res. Foundation, Denver Co, 256-281.

Tang, H. and Shi, B. (2002) The characteristics of composite flocculants synthesized with inorganic polyaluminium and inorganic polymers. In: *Chemical Water and Wastewater Treatment VII* (Eds: Hahn, H.H., Hoffmann, E and Ødegaard, H.), pp 17-28, IWA Publishing, London.

Tang, H. X. and Luan, Z. K. (1996) The differences of behaviour and coagulating mechanism between inorganic polymer flocculants and traditional coagulants. In: *Chemical Water and Wastewater Treatment IV* (Eds: Hahn, H.H., Hoffmann, E and Ødegaard, H.), pp 83-93, Springer-Verlag, Berlin.

Tang, H. X., Luan, Z. K., Wang, D. S. and Gao, B. Y. (1998) Composite inorganic flocculants. In: *Chemical Water and Wastewater Treatment V* (Eds: Hahn, H.H., Hoffmann, E and Ødegaard, H.), pp 25-34, Springer-Verlag, Berlin.

Tang, H. X., Wang, D. S. and Xu, Y. (2004) Optimisation of the concepts for poly-aluminium species. In: *Chemical Water and Wastewater Treatment VIII* (Eds: Hahn, H.H., Hoffmann, E and Ødegaard, H.), pp 139-149, IWA Publishing, London.

Tchoubar, D., Bottero, J. Y., Quienne, P. and Arnaud, M. (1991) Partial hydrolysis of ferric chloride salt. Structural investigation by photon-correlation spectroscopy and small angle X-ray scattering. *Langmuir*, 7, 398-402.

TeKippe, R. J. and Ham, R. K. (1971) Velocity-gradient paths in coagulation. *J. AWWA.*, 63, 429-448.

Tomi, D.T. and Bagster, D.F. (1978) The behaviour of aggregates in stirred vessels. Part I- Theoretical considerations on the effects of agitation. *Trans. Inst. Chem. Eng*, 56, 1-8.

Valade, M. T., Edzwald, J. K., Tobiason, J. E., Dahlquist, J., Hedberg, T. and Amato, T. (1996) Particle removal by flotation and filtration: pretreatment effects. *J. AWWA*, 88(12), 35-46.

Van Benschoten, J.E. and Edzwald, J.K. (1990a) Chemical aspects of coagulation using aluminium salts – I. Hydrolytic reactions of alum and polyaluminium chloride. *Water Res.*, 24 (12), 1519-1526.

Van Benschoten, J.E. and Edzwald, J.K. (1990b) Chemical aspects of coagulation using aluminium salts – II. Coagulation of fulvic acid using alum and polyaluminium chloride. *Water Res.*, 24, 1527-1535.

Wakeman, R. J. (1981) Thickening and filtration: a review and evaluation of recent research. *Trans. Inst. Chem. Eng.*, 59, 147-160.

Warden, J. H. (1983) *Sludge Treatment Plant for Waterworks*. Tech. Rep. TR189, Water Research Centre, Swindon, UK.

Watts M C. (1994) *Revised Guidelines for the Rapid Dispersion of Chemicals in Water Treatment.* Rep. No. UC 2224, WRc, Swindon, UK.

Webster, J. A., Webster, H. C. and Fairley, J. P. (1977) Aspects of design for efficient plant operation for water treatment using coagulants. Unpublished paper presented at joint meeting of Scottish branches of IWES, IPHE & IWPC, Glasgow, Feb. 1977.

Wiesner, M. R. (1992) Kinetics of aggregate formation in rapid mix. *Water Res.*, 26(3), 379-387.

Williams, R. A., Peng, S. J. and Naylor, A. (1992) In situ measurement of particle aggregation and breakage kinetics in a concentrated suspension. *Powder Technol.*, 77(1), 75-83.

Wu, C.C., Huang, C. and Lee, D. J. (1997) Effects of polymer dosage on alum sludge dewatering characteristics and physical properties. *Colloids Surfaces A*, 122, 89-96.

Wu, R. M. and Lee, D. J. (1998) Hydrodynamic drag force exerted on a moving floc and its implications to free-settling tests. *Water Res.*, 12(3), 760-768.

Yao, K. M., Habibian, M. T. and O'Melia, C. R. (1971) Water and wastewater filtration: concepts and applications. *Environ. Sci. Tech.*, 5(11), 1105-1112.

Yadav, N.P. and West, J.T. (1975) *The Effect of Delay Time on Floc Blanket Efficiency.* Rep. No. TR9, WRc, Swindon, UK.

Yeung, A. K. C. and Pelton, R. (1996) Micromechanics: A new approach for studying the strength and breakup of flocs. *J. Colloid Interface Sci.*, 184(2), 597-585.

Zhanpeng, J. and Yuntao, G. (2006) Flocculation morphology-effect of particulate shape and coagulant species on flocculation. *Water Sci. Tech.*, 53(7), 9-16.

Zhao, Y. Q. and Bache, D. H. (2002) Polymer impact on filter blinding during alum sludge filtration. *Water Res.*, 36, 3691-3698.

Zhao, Y. Q., Papavasilopoulos, E. N. and Bache, D. H. (1998) Clogging of filter medium by excess polymer during alum sludge filtration. *Filt. Sep.*, 35(10), 947-950.

Appendix A
SI Units and Physical Constants

SI BASE UNITS

Of the seven dimensionally independent base units, the ampere (A) is not used in the book.

Quantity	*Name*	*Symbol*
Mass	kilogram	kg
Length	metre	m
Time	second	s
Temperature	kelvin	K
Amount of substance	mole	mol
Electric current	ampere	A
Luminous intensity	candela	cd

SI DERIVED UNITS

Listed are some named derived units used in the book:

Quantity	*Name*	*Symbol*	*Definition*
Force	newton	N	$kg\ m\ s^{-2}$
Pressure	pascal	Pa	$N\ m^{-2} \equiv J\ m^{-3} \equiv kg\ m^{-1}\ s^{-2}$
Energy	joule	J	$N\ m \equiv kg\ m^{2}\ s^{-2}$
Power	watt	W	$J\ s^{-1} \equiv kg\ m^{2}\ s^{-3}$
Electric charge	coulomb	C	A s
Electric potential difference	volt	V	$J\ C^{-1} \equiv kg\ m^{2}\ s^{-3}\ A^{-1}$
Electric capacitance	farad	F	$C\ V^{-1} \equiv kg^{-1}\ m^{-2}\ s^{4}\ A^{2}$
Frequency	hertz	Hz	s^{-1}

SI PREFIXES

Listed are prefixes to designate submultiples from to 10^{-9} to 10^{9} of a base unit:

Factor	*Prefix*	*Symbol*	*Factor*	*Prefix*	*Symbol*
10^{-1}	deci	d	10	deca	da
10^{-2}	centi	c	10^{2}	hecto	h
10^{-3}	milli	m	10^{3}	kilo	k
10^{-6}	micro	μ	10^{6}	mega	M
10^{-9}	nano	n	10^{9}	giga	g

NON-SI UNITS

Confined to those used in text:

Quantity	*Name*	*Symbol*	*Definition*
Atomic mass unit (= 1/12 Mass of carbon-12)	dalton	Da	1.6605×10^{-27} kg
Colour	hazen	°H	≡ colour corresponding to 1mg/l Pt/Co
Concentration	molar	M	mol dm^{-3}
Length	angstrom	Å	10^{-10} m
Mass	tonne	t	10^{3} kg
Pressure	bar	bar	10^{5} Pa
Temperature	degree Celsius	°C	= T(K) – 273.15
Volume	litre	*l* or l	10^{-3} m^{3} ≡ dm^{3}
Time	hour	h	3600 s
Turbidity	nephelometric turbidity unit	NTU	Optical density based on Formazin concentration

PHYSICAL CONSTANTS

Quantity	*Symbol*	*Value*	*SI Units*
Avogadro number	N_A	6.022×10^{23}	mol^{-1}
Boltzmann constant	k	1.381×10^{-23}	J K^{-1}
Elementary charge	e	1.602×10^{-19}	C
Faraday constant	F	9.648×10^{4}	C mol^{-1}
Gas constant	R	8.314	J K^{-1} m^{-1}
Permittivity of vacuum	ε_o	8.854×10^{-12}	C V^{-1} m^{-1}
Standard acceleration of gravity	g	9.807	m s^{-2}

Appendix B
Symbols

All standard mathematical and chemical symbols are taken to have their usual meaning. In places, bold face vector notation has been employed.

Many symbols are defined locally and no attempt is made to define every symbol in this listing.

The book encompasses many disciplines, so in trying to use standard notation within the disciplines, repetition of symbols has been unavoidable.

An overbar refers to a mean value (usually with respect to time).

Primes usually denote departures from a mean value; an exception is A′.

Symbol	*Definition*	*Definition or appearance*	*Units*
a	Fitting coefficient in gamma distribution	Equation 2.151	
a	Radius of sphere	Equation 1.2	m
A	Area		m^2
A_c	Projected cross sectional area	Equation 2.61	m^2
A_H	Hamaker constant	Equation 1.2	J
A′	Packing factor or pre-factor	Equation 2.14	
b	Fitting coefficient in gamma distribution	Equation 2.151	
$\|b(0)\|^2$	Scattering cross-section of particle	Equation 3.7	m^2
$b(\mathbf{q})$	Scattering length of particle	Equation 3.6	m
B	Fraction of moisture removed in secondary consolidation	Section 3.2.7.10	
B	Structure pre-factor	Equation 2.96	
c	Fitting coefficient in gamma distribution	Equation 2.151	
c, C	Mass concentration	Section 1.4.4.1	$kg\ m^{-3}$
c, c_R	Volume average solids concentration within cluster	Equation 2.22	$kg\ m^{-3}$
c′	Concentration fluctuation in turbulent field	Section 1.4.4.1	$kg\ m^{-3}$
C	'Strength' constant in $d_{max} = C/G^m$	Equation 2.112	$m\ s^{-m}$
CD	Charge density (mol% of ionogenic groups); Also expressed as:	Section 1.6	mol% $C\ g^{-1}$ $meq\ g^{-1}$
C_D	Drag coefficient	Equation 2.61	

d	Spherical equivalent diameter		m
d	Space dimension; *e.g.* = 1, 2, 3,…	Section 2.3.6.1	
d_b	Diameter of bubble	Section 3.3.4	m
d_g	Spherical equivalent grain diameter	Section 3.3.6.1	m
d_L	Arithmetic mean diameter of number size distribution	Section 2.5.2	m
d_{pb}	Spherical equivalent diameter of particle-bubble cluster	Equation 3.32	m
d_{max}	Spherical equivalent diameter of largest flocs	Equation 2.112	m
d_{L50}	50th percentile in length size distribution	Section 3.2.7.3	m
d_{V95}	95th percentile in volume size distribution	Section 2.5.3	m
D	Fractal dimension (in much of the text, D refers to the *mass fractal dimension*)	Equation 2.14	
D_c	Molecular diffusion coefficient of tracer	Section 1.4.4.2	$m^2\ s^{-1}$
D_q	Rényi dimension (q = 0,1,2,…)	Section 2.1.2.4	
D_s	Fractal dimension of single bonded mesh elements within gel	Section 2.3.3	
D_B	Box-counting dimension ($\equiv D_0$)	Section 2.1.2.1	
D_C	Correlation dimension ($\equiv D_2$)	Section 2.1.2.3	
D_I	Information dimension ($\equiv D_1$)	Section 2.2.2.2	
e	Elementary charge	Appendix A	C
e	Strain (= relative change in length)	Section 2.3.1	
E	Aspect ratio	Section 2.2.3.2	
E	Mechanical energy per unit volume required to break backbone in mesh	Section 2.3.5.1	$J\ m^{-3}$
E	Strain rate	Figure 1.18	s^{-1}
E	Young's modulus	Section 2.3.1	Pa
$\bar{E}$	Energy per unit volume dissipated when mesh elements re-established	Section 2.3.5.1	$J\ m^{-3}$
$E(k)$	Energy spectral density	Section 1.4.2.3	$m^3\ s^{-2}$
$E_c(k)$	Spectral density of concentration fluctuations	Equation 1.40	$kg^2\ m^{-5}$
E_o	Amplitude of electric field	Section 3.1.3.3	$V\ m^{-1}$
E_f	Free energy quantity	Section 2.3.4	J
E_{sep}	Separation energy per contact	Equation 2.94	$J\ m^{-3}$
f	Fraction		
f_S	Shape factor	Equation 2.64	
f_T	Translational friction coefficient	Equation 2.45	$N\ m^{-1}\ s$
F	Bond strength	Equation 1.19	N
F	Force		N
F	Faraday constant; = $N_A e$	Equation 1.4	$C\ mol^{-1}$
F	Theoretical filter capacity; = $c\varepsilon_o$	Section 3.3.6.3	$kg\ m^{-3}$
F_D	Drag force	Equation 2.61	N

F_{gn}	Net gravitational force	Section 2.2.1	N
F_θ	Solids concentration in deposit; = $c\sigma_v$	Equation 3.46	kg m^{-3}
g	Gravitational acceleration	Equation 1.40	m s^{-2}
G	Mean velocity gradient	Equation 1.39	s^{-1}
G	Shear modulus	Equation 2.75	Pa
G_o	Modulus of fast elastic deformation	Equation 2.76	Pa
G_p	Shear modulus of particle	Equation 2.86	Pa
G_p	Structure function of order p	Equation 2.122	m^{2p} s^{-2p}
$G^*(\omega)$	Dynamic shear modulus	Section 2.3.4	Pa
h	Head loss	Equation 1.40	m
h	Shortest distance of separation between particle surfaces	Equation 1.2	m
h_c	Depth of filter cake	Equation 3.59	m
I	Information entropy	Section 2.1.2.2	
I	Ionic strength	Section 1.2.3	mol m^{-3}
I_q	Entropy of order q	Section 2.1.2.5	
J	Collision frequency per unit volume	Equation 1.10	m^3 s^{-1}
k	Collision rate constant	Equation 1.10	m^3 s^{-1}
k	Boltzmann constant	Section 1.2.10	J K^{-1}
k	Mass transfer coefficient	Equation 3.19	m^3 s^{-1}
k	Permeability of porous medium	Equation 2.36	m^2
k	Wave number	Section 1.4.2.3	m^{-1}
k	Propagation vector	Section 3.1.3.3	m^{-1}
k_o	Orthokinetic flocculation rate constant	Equation 1.18	m^3 s^{-1}
k_o	Wave number of maximum in E(k)	Figure 1.19	m^{-1}
k_p	Perikinetic flocculation rate constant	Equation 1.12	m^3 s^{-1}
k_B	Batchelor wave number; = $(\varepsilon/\nu D_c^2)^{1/4}$	Section 1.4.4.2	m^{-1}
k_R	Internal permeability of aggregate	Equation 2.27	m^2
k_R	Rupture rate constant	Equation 1.24	s^{-1}
k_η	Wave number corresponding to Kolmogorov length	Figure 1.19	m^{-1}
K	Dimensionless permeability	Equation 2.47	
l	Eddy size scale	Section 1.4.2.1	m
l	Length		m
l	Spacing distance	Section 2.4.3	m
L	Bed depth	Equation 3.22	m
L_c	Integral length scale of mass concentration field	Section 1.4.4.2	m
L_c	Length of contact zone in DAF unit	Equation 3.30	m
L_F	Characteristic length in elastic network associated with applied force F	Section 2.3.6.1	m
m	Index value; see also Equation 2.141	Equation 2.112	
m	Mass		kg
m_{bw}	Mass of bound water	Section 2.1.5	kg
m_{fw}	Mass of free water	Section 2.1.5	kg
m_m	Mass of solid-water mixture	Section 2.1.3.1	kg
m_s	Mass of solids	Section 3.3.7.2	kg

m_{sl}	Mass of solid-liquid mixture (sludge)	Section 3.3.7.2	kg
m_w	Mass of water	Section 3.3.7.2	kg
M	Molar mass	Section 1.2.10	kg mol^{-1}
M	Molecular weight		Da
M	Suspended solids concentration	Equation 2.148	kg m^{-3}
n (d)	Number of flocs of size d per unit volume	Section 2.5.2	m^{-3}
n	Ligand ratio; OH/Me in compound	Section 1.5.3.3	
n	Particle concentration	Section 1.3.3	m^{-3}
n_b	Number of bonds broken	Equation 2.130	
n_b	Number concentration of bubbles	Equation 3.28	m^{-3}
n_c	Number concentration of collectors	Equation 3.19	m^{-3}
n_p	Number concentration of incoming particles/flocs towards collectors	Equation 3.19	m^{-3}
N	Number of contacts between particles in aggregate	Equation 2.94	
N	Number of flocs per unit volume	Equation 2.145	m^{-3}
N	Number of primary particles in cluster	Equation 2.14	
N	Rotation rate	Section 1.4.4.1	s^{-1}
N_A	Avogadro constant	Section 1.6	mol^{-1}
N_s	Number of particles in single bonding pathway in mesh length ξ	Equation 2.83	
p	Order of structure function G_p	Section 2.4.3.2	
p	Order of distribution moment	Equation 2.157	
P(q)	Form factor of scattering particle	Section 3.1.3.3	
P	Pressure	Equation 2.36	Pa
P	Power input	Section 1.4.3.1	J s^{-1}
P	Fraction of broken strands in network	Section 2.3.5	
P_c	Critical fraction of broken strands in network (when $P<P_c$ network is continuous)	Section 2.3.5	
P_i	Probability of finding point in cell	Equation 2.4	
P_L	Effective liquid pressure	Figure 3.31	Pa
P_s	Particle pressure (transmitted stress)	Equation 2.106	Pa
P_y	Compressive yield stress	Equation 2.106	Pa
q	Charge strength	Equation 1.9	C
q	Magnitude of scattering vector **q**	Equation 3.5	m^{-1}
q	Moment of probability	Section 2.1.2.5	
q	Scattering vector	Figure 3.6	m^{-1}
Q	Volumetric flow rate		m^3 s^{-1}
r	Weighting coefficient; $\sigma \propto \phi^r$	Equation 2.134	
r	Hydrodynamic radius	Section 2.2.2	m
r	Radius or radial coordinate		m
r_{OH}	Molar ratio [OH]/[Me_T] in solution	Section 1.5.3.2	
R	Distance from reference origin	Section 3.1.3.3	m
R	Gas constant	Equation 1.4	J K^{-1} mol^{-1}
R	Hydraulic resistance	Equation 3.57	m^{-1}

R	Radius of spherical cluster	Equation 2.14	m
R	Radius of pore	Equation 2.114	m
R_c	Specific resistance of filter cake	Equation 3.50	$m\ kg^{-1}$
R_f	Rate of fast aggregation	Section 1.3.3.2	$m^3\ s^{-1}$
R_m	Hydraulic resistance of filter membrane	Equation 3.58	m^{-1}
R_s	Rate of slow aggregation	Equation 1.16	$m^3\ s^{-1}$
Re	Reynolds number (*e.g.*= ud/ν)	Equation 2.62	
R_G	Radius of gyration	Equation 1.53	m
S	Cluster 'strength'	Equation 2.129	Pa
S	Coefficient of compressibility	Equation 3.62	
S	Entropy	Equation 2.84	$J\ K^{-1}\ m^{-3}$
S	Surface area of ellipsoid	Equation 2.118	m^2
S_o	Cluster 'strength' scaling factor; $r = 1$	Section 2.4.4	Pa
S_o'	Cluster 'strength' scaling factor; r general	Section 2.4.4	Pa
Sc	Schmidt number; $=\nu/D_c$	Section 1.4.4.2	
S_f	Structure factor; $\equiv A'$ (when associated with permeable aggregates)	Section 2.2.2	
$S(D)$	Distribution moment of order D conveying information about distribution of mass	Equation 2.155	
$S(q)$	Static structure factor of scattering system	Section 3.1.3.3	
t	Time		s
t_c	Turnover time ($\equiv$ periodicity)	Section 1.4.4	s
t_L	Retention time in collector system	Equation 3.25	s
t_r	Recovery timescale in elastic network	Equation 2.90	s
T	Temperature		K or °C
u	Superficial velocity; $=Q/A$	Equation 2.36	$m\ s^{-1}$
u	Velocity		$m\ s^{-1}$
u	Scaled floc size; $=d/d_L$	Section 2.5.2	
u'	Velocity fluctuation in turbulent field	Section 1.4.4.2	$m\ s^{-1}$
u^2	Standard notation for variance of u' [time average value of $(u')^2$]	Equation 1.35	$m^2\ s^{-2}$
u_o	Settling of isolated sphere in viscous flow ($Re < 0.5$)	Equation 2.58	$m\ s^{-1}$
u_0	Velocity scale	Section 2.4.2.3	$m\ s^{-1}$
u_f	Interstitial (or pore) velocity	Equation 2.37	$m\ s^{-1}$
u_s	Average velocity of solids	Equation 2.41	$m\ s^{-1}$
U	Internal energy	Equation 2.84	$J\ m^{-3}$
U	Permeability scaling factor	Section 2.2.2	
U_c	Consolidation ratio	Equation 3.64	
v_o	Scaling velocity	Equation 3.21	$m\ s^{-1}$
v_b	Rise rate of bubble relative to surroundings	Equation 3.28	$m\ s^{-1}$
v_d	Velocity difference	Section 2.4.3.3	$m\ s^{-1}$
v_K	Kolmogorov velocity scale	Equation 1.3.3	$m\ s^{-1}$

v_{pb}	Rise rate of particle-bubble system	Section 3.3.5	$m\ s^{-1}$
V	Volume		m^3
V_A	Attraction potential associated with van der Waals interaction between particles	Section 1.2.2	J
V_B	Potential associated with Born forces	Equation 1.1	J
V_{cell}	Cell volume	Section 2.2.1	m^3
V_R	Repulsive potential energy between two charged particles	Section 1.2.4	J
V_S	Structural energy between particles	Section 1.2.1	
v_{sett}	Settling velocity in hindered settling	Equation 3.21	$m\ s^{-1}$
V_{sl}	Volume of solid-liquid mixture (sludge)	Equation 3.50	m^3
V_T	Total potential energy	Equation 1.1	J
VAD	Volume average diameter	Equation 2.159	m
w	Fluid velocity, z component		$m\ s^{-1}$
W	Stability ratio	Equation 1.15	
x	Cartesian coordinate		m
x_p	Hydrodynamic size of polymer	Equation 1.53	m
y	Cartesian coordinate		m
z	Cartesian coordinate normal to surface		m
α	Collision efficiency factor	Equation 1.11	
α	Coefficient in incomplete gamma function	Equation 2.160	
α	Fitting coefficient	Equation 3.66	m^{-1}
α	Lipschitz-Holder exponent	Figure 2.2	
α	Index value ; = $(3+D_s)/(3-D)$	Equation 2.86	
α_{pb}	Particle-bubble collision efficiency	Section 3.3.5	
α_{pc}	Particle-collector collision efficiency	Section 3.3.1.2	
β	Fitting coefficient	Equation 3.63	m^{-3}
β	Inverse permeability ratio; $=R/\sqrt{k}$	Equation 2.70	
β	Index value; = $D_s/(3-D)$	Equation 2.108	
β	Proportionality coefficient	Equation 2.130	
β	Proportionality coefficient	Equation 3.59	
β_p	Universal function of order p	Section 2.4.3.2	
γ	Deformation (or strain)	Figure 2.8a	
γ	'Strength' coefficient; see Equation 2.140	Equation 2.139	$m^{1+m}\ s^{-3m/2}$
γ_o	Fast elastic deformation	Section 2.3.1	
γ_c	Critical deformation; = ξ/R	Section 2.3.4	
$\dot{\gamma}$	Deformation rate (or strain rate)	Equation 2.77	
δ	Index value; = $3/(3-D)$	Equation 2.93	
δ	Probing length	Equation 2.1	m
δ	Thickness of Stern layer	Section 1.2.3	m
δ_0	Reference length scale	Equation 2.26	m
δ_ω	Size of smallest stable vortex	Section 1.4.2.2	m
ΔP	Pressure difference		Pa

Δu	Velocity difference	Equation 2.122	$m\ s^{-1}$
$\Delta \rho$	Density difference	Equation 2.60	$kg\ m^{-3}$
ε	Porosity		
ε	Rate of energy dissipation per unit mass	Section 1.3.3.3	$m^2\ s^{-3}$
$\bar{\varepsilon}$	Volume average value of rate of energy dissipation per unit mass	Equation 1.39	$m^2\ s^{-3}$
ε	Permittivity of a medium	Section 1.2.3	$C^2\ N^{-1}\ m^{-2}$
ε_o	Permittivity of a vacuum	Section 1.2.3	$C^2\ N^{-1}\ m^{-2}$
ε_o	Porosity of clean bed	Equation 3.37	
ε_c	Porosity of filter cake	Equation 3.53	
ε_r	Relative permittivity or Dielectric constant	Section 1.2.3	
ε_R	Volume average of cluster porosity	Equation 2.21	
ζ	Index value; $= (3+2D_s)/(3-D)$	Equation 2.109	
ζ	Zeta potential	Section 1.2.3	V
η	Kolmogorov length	Equation 1.31	m
η	Creep coefficient in secondary consolidation	Equation 3.65	s^{-1}
η_{pb}	Particle-bubble attachment efficiency	Section 3.3.5	
η_{pc}	Particle-collector attachment efficiency	Section 3.3.1.2	
η_{rapid}	Value of η associated with rapid mixing	Section 3.2.7.1	m
θ	Deposit thickness	Equation 3.42	m
θ	Moisture content	Section 2.1.5	
θ	Angle		
θ_c	Consolidation time	Section 3.2.7.10	s
κ	Debye-Huckle parameter	Section 1.2.3	m^{-1}
Λ	Macro length scale of turbulence	Section 1.4.2.1	m
λ	Filter coefficient	Equation 3.38	m^{-1}
λ_o	Filter coefficient of clean bed	Equation 3.39	m^{-1}
λ	Taylor length scale	Equation 1.30	m
λ	Wavelength of electromagnetic radiation	Section 3.1.3.3	m
λ_l	Relative longitudinal deformation	Equation 2.105	
λ_F	Relative deformation	Equation 2.101	
μ	Dynamic viscosity	Equation 1.12	Pa s
μ_m	Viscosity of fluid medium	Equation 2.111	Pa s
μ_o	Creep viscosity (at low strain rate)	Figure 2.10	Pa s
μ_∞	Limiting viscosity (at high shear)	Figure 2.10	Pa s
μ_p	Plastic viscosity	Equation 2.78	Pa s
ν	Kinematic viscosity; $=\mu/\rho_{fl}$	Section 1.3.3.3	$m^2\ s^{-1}$
ν_P	Poisson ratio	Equation 2.100	
ξ	Mesh size (or correlation length)	Section 2.3.3	m
ξ	Effective cluster diameter in porous cake	Equation 3.54	m

ρ	Density; wider concept than kg m^{-3}	Equation 2.16	
ρ_d	Density of accumulated deposit in filter	Equation 3.41	kg m^{-3}
ρ_e	Effective density; = ρ_s - ρ_{fl}	Equation 2.24	kg m^{-3}
ρ_f	Average density of floc	Section 3.3.4	kg m^{-3}
Φ_F	Volume fraction of flocs	Equation 3.17	
ρ_{fl}	Density of suspending fluid	Equation 2.43	kg m^{-3}
ρ_m	Density of mixture *e.g.* solids and water	Equation 2.25	kg m^{-3}
ρ_s	Density of solids	Equation 2.25	kg m^{-3}
ρ_w	Density of water	Equation 1.59	kg m^{-3}
σ	Shear stress	Section 2.3.1	Pa
σ	Tensile stress	Section 2.3.1	Pa
σ_o	Stress scaling factor	Equation 2.134	Pa
σ_B	Bingham yield stress	Equation 2.78	Pa
σ_t	Turbulent stress	Equation 2.120	Pa
σ_v	Specific deposit	Equation 3.37	
σ_{vu}	Ultimate specific deposit	Section 3.3.6.3	
σ_y	True yield stress	Figure 2.11	Pa
σ_σ	Charge density at plane of shear	Equation 1.6	C m^{-2}
τ	Timescale		s
τ_K	Kolmogorov time scale	Equation 1.34	s
τ_s	Stress relaxation time for solid-like state	Section 2.3.7	s
φ	Energy associated with separation	Equation 2.127	J
ϕ	Solids volume fraction	Equation 1.20	
$\phi(r)$	Local value of ϕ at cluster radius r	Equation 2.16	
ϕ_o	Cluster volume fraction scaling factor	Equation 2.18	
ϕ_o	Solids volume fraction of primary particles	Equation 2.49	
ϕ_c	Solids volume concentration in consolidation layer or filter cake	Figure 3.36 & Section 3.3.7.9	
ϕ_F	Volume fraction of aggregates	Equation 2.49	
ϕ^*	Critical value of ϕ_F at cluster overlap ($\cong \phi_g$)	Equation 2.79	
ϕ_g	Value of ϕ at gel point	Section 2.3.3	
ϕ_R	Volume average of ϕ in cluster radius R	Equation 2.15	
ϕ_ξ	Volume average of ϕ_o in cluster diameter ξ	Equation 3.54	
Φ_r	Relative flux of alkali neutralisation	Equation 1.49	s^{-1}
ψ	Electric potential	Equation 1.3	V
ψ	Water potential	Section 3.3.7.2	Pa
ψ_δ	Electric potential at edge of Stern layer	Equation 1.3	V
$\Psi(u)$	Number size distribution in scaled form	Section 2.5.2	
ω	Frequency of oscillation	Section 2.3.4	s^{-1}
Ω	Relative drag resistance	Equation 2.65	

Subscripts

i	With reference to ith species
j	With refererence to jth species
n	With reference to nth species
$_{11}$	Component in stress tensor σ_{11}; component in rate of strain tensor γ_{11}
∞	Limiting value when variable (*e.g.* time or shear stress) tends to infinity

Superscript

a	Oligomer component in timed calorimetric analysis; Me^a
b	Polymeric component in timed calorimetric analysis; Me^b
c	Colloidal component in timed calorimetric analysis; Me^c
p	Order of distribution moment
$'$	Perturbation component; with exception of A'

Mathematical

$\langle\ldots\rangle$	Average value; usually with respect to time, but may include localised volume within which parameter defined; also refers to average across a distribution
lim	Limiting value

Acronyms/nomenclature

ACN	Adsorption-charge neutralisation
ALT	Aluminium to turbidity ratio
am	Amorphous; as in $Al(OH)_3(am)$
AVI	Aggregated volume index
AWWA	American Water Works Association
BET	Brunauer-Emmett-Teller; as in BET model for adsorption
BW	Bound water
c.c.c.	Critical coagulant concentration
CCTV	Closed circuit television
CD	Charge density
CML30	Name of settlement test; 30 min settling in 100 ml measuring cylinder
CST	Capillary suction time
DAF	Dissolved air flotation
DHB	Author
DLA	Diffusion limited aggregation
DOC	Dissolved organic carbon
DLVO	Derjaguin-Landau-Verwey-Overbeek ; as in DLVO theory
DOM	Dissolved organic matter
EOM	Extracellular organic matter
FBC	Floc blanket clarifier
FI	Flocculation index
HPAM	Hydrolysed polyacrylamide

LED	Light-emitting diode
MC	Moisture content
Me	Metal
MW	Molecular weight
NOM	Natural organic matter
NTU	Nephelometric turbidity unit
PAA	Polyacrylic acid
PAC	Powdered activated carbon
PACl	Polyaluminium chloride
PAFS	Poly-alumino-iron-sulphate
PAM	Polyacrylamide
PASS	Polyaluminium-silicate-sulphate
PAX	A trade name for polyaluminium compounds
PDADMAC	Cationic polydiallyldimethyl ammonium chloride
PFSS	Polyferric-silicate-sulphate
ppt	Precipitate
PSD	Particle size detector
RG	Author
RGD	Raleigh-Gans-Debye; as in RGD approximation
RLA	Reaction limited aggregation
SRF	Specific resistance to filtration
SUVA	Ultraviolet absorption at 256nm per unit of TOC
THM	Trihalomethane
TOC	Total organic carbon
v/v	Volume ratio
WTW	Water treatment works
w/w	Mass ratio

Acknowledgements

We acknowledge the following and thank the quoted publishers and individuals for granting permission to reproduce material; full references are listed at the end of the principal sections.

American Chemical Society, Washington D.C.
Figure 1.27
American Institute of Chemical Engineers, New York
Figure 1.15
American Society of Civil Engineers, Reston VA
Figure 3.32
American Water Works Association, Denver Co
Figures: 1.25, 1.28; Tables: 3.20, 3.27, 3.28
Chartered Institution of Water and Environmental Management, London
Table 3.23
Elsevier, Oxford
Figures: 1.1, 1.4, 1.18, 1.23, 1.24, 1.26, 1.29, 2.7, 2.11, 2.15, 2.18, 3.8, 3.10, 3.18, 3.40-3.43; Tables: 1.7, 2.2, 3.18
Institution of Chemical Engineers, London
Figure 3.9
IWA Publishing, London
Figures: 1.8-1.10, 1.13, 1.14, 2.4, 3.15-3.17, 3.19, 3.22-3.25, 3.36, 3.38
Tables: 3.5, 3.6, 3.11, 3.12
McGraw-Hill Education, New York
Figure 1.3
Springer, Heidelberg
Figures: 2.8-2.10, 3.2, 3.4, 3.13, 3.27; Table 1.4
Taylor & Francis,
Figures: 1.12, 2.6
Wiley, Hoboken NJ
Figures 1.32, 2.1
WRc, UK
Tables: 1.1 (via T Hall), 3.8, 3.9, 3.13 (via G Dillon and Associates)
Individual
Professor Mark M Clark, University of Illinois; Figure 1.21

Index

ISBN: 9781789065442. Published by IWA Publishing, London, UK.